Springer Praxis Books

More information about this series at http://www.springer.com/series/4097

Linda Dawson

The Politics and Perils of Space Exploration

Who Will Compete, Who Will Dominate?

Linda Dawson
Interdisciplinary Arts and Sciences
University of Washington
Tacoma, WA, USA

Springer Praxis Books
ISBN 978-3-319-38811-3 ISBN 978-3-319-38813-7 (eBook)
DOI 10.1007/978-3-319-38813-7

Library of Congress Control Number: 2016948726

Printed on acid-free paper

This Springer imprint is published by Springer Nature
The registered company is Springer International Publishing AG
The registered company address is: Gewerbestrasse 11, 6330 Cham, Switzerland

To my family - Mom and Dad, my sisters Judi and Patti, and my husband Allan.

Preface

One day I asked my college students (not science majors) how their cell phone transmits and receives data. Several students knew that satellites were involved but didn't understand the underlying concepts of geosynchronous orbits and how objects can stay in position over the same region as they circle Earth. In my History and Science of Space Exploration class, I had a student who thought Lance Armstrong was the first man on the Moon, as well as someone else who defined the concept of a vacuum in space as somehow involving sucking air out of an object, Hoover style.

Maybe they were just joking, but it's not big news that students aren't aware of how science works or affects our everyday life. It is, however, disappointing to me that students have so little knowledge of outer space, satellites, our Solar System, and a basic knowledge of the history of space exploration. The students were unaware of a human presence in the International Space Station, how they get there, why they are there, and if NASA still existed. In addition, we were clearly talking about the "history" of space exploration. It was no longer current news that inspired them, although many were enthusiastic about deep space exploration, fueled by science fiction, both in books and movies. I was the only one in the classroom who had actually lived through and experienced the space race up through the space shuttle and the International Space Station. Being aware enough to experience human space exploration from its start to current day has given me a unique perspective.

I wanted to write this book for a number of reasons. I felt that if my students were confused about the existence of NASA today and its relationship to private space enterprises, then others were most likely equally confused. In addition, I wanted to write about the one subject that still inspires me. Through all of the education and jobs I have had, I have always loved airplanes, rockets, and the space program. The underlying focus of my career has been science and technology. A visit to Cape Canaveral early in college inspired me to study space exploration through aerospace engineering. The Moon race fueled my enthusiasm and helped me succeed through the difficult engineering curriculum at MIT, George Washington University, and the University of Washington. I have always kept a connection with space. This project is one more way of exploring more about the interconnections of international space exploration and keeping the subject fresh and alive in my life.

My goal for this book is to provide some insight into current efforts in space exploration, primarily manned efforts. During the space race, the players and the objectives were straightforward, but today, dozens of countries have participated in some aspect of space. Many countries are interested in scientific exploration, security, and communications. Only a few can support the high price of manned missions, which have been confined to low-Earth orbit since the last Moon mission. The relationships between the space programs and their goals are varied and often determined by the resources available. The complexities can often be simplified by looking at regional goals rather than worldwide ambitions. Shared objectives can be combined in partnership efforts. As in other aspects of combined efforts, power and achievement can be accomplished by joining forces and available resources.

Another objective in writing this book is to address how politics has affected the direction of the American space program. Whether it is international or local politics, it is clear that government programs follow government priorities and their associated funding. Success of the Apollo program was one of the most important priorities for U.S. government funding in the 1960s. Our national and international priorities have gone through major changes. Recent focus on military and wartime efforts in the Middle East has depleted the available budget for space efforts. The dependence of NASA and the space program on government funding resulted in limited resources devoted to space missions and space science research. Other types of funding and public support were needed to support more robust space activities. Because of strong public interest and entrepreneurs that were visionaries, a number of small, medium, and large space-related businesses started development on a variety of systems that are crucial to space travel taking the next step.

My career in aerospace engineering includes working at NASA in Houston on the space shuttle program for years prior to the first launch and past the first couple of launches. I was hired to be an Aeronautical Flight Controller for Mission Control. This is the first vehicle that would operate as an airplane on re-entry, requiring the development of a series of operational tools. The space shuttle orbiter vehicle was already designed, developed, and being built in the mid- to late 1970s. When I was hired by NASA, the prototype *Enterprise* was about to be transported and drop-tested from a 747 airplane to test its glide capabilities. I became familiar with the shuttle vehicle and NASA operations and came to understand how stable the orbiter vehicle would be during its re-entry maneuvers. After initiating a de-orbit burn, the orbiter would go through a series of S-turns designed to slow the spacecraft down prior to landing. No other combination airplane/spacecraft had flown before at hypersonic speeds outside of Earth's atmosphere. There were a lot of unknowns. My group investigated other hypersonic aircraft such as the SR-71, the X-15, and experimental lifting bodies to gain insight into the behavior of the orbiter as a glider. As it turned out, the orbiter vehicle was very stable and never became unstable in its descent.

One of my major tasks at NASA was to help develop the flight rules for the orbiter primarily for entry operations in addition to an abort re-entry. The development of these rules required participation in extensive simulations for de-orbit and re-entry. I developed and conducted some of these studies using a re-entry simulator flown by shuttle astronauts. Another component of my job was to estimate how much fuel was necessary to control the vehicle in case of stability problems. After the de-orbit burn, the only control for the orbiter vehicle comes from small reaction control jets that were used for orbital maneuvering and control during entry or orbit maneuvers in the highest part of the atmosphere. If a control jet fails or another control problem requires a jet to stay on or off, vehicle control is maintained by the opposite reaction control jets, staying on to compensate and maintain control. This type of failure uses an extra quantity of fuel. To conserve fuel and save weight, only so many of these malfunctions can be accommodated. Therefore, the failures are prioritized as the most or least likely. After extensive simulations, the final entry fuel budget reflected my simulation study for entry failures.

The final component to my training as a flight controller involved a series of extensive integrated simulations prior to the first shuttle launch. The purpose of these simulations was to verify all systems were being monitored properly and if one or more systems failed, that the appropriate steps were taken to assist the crew and fix the problem. The truth is, for launch and re-entry, there is little that flight controllers could recommend that would be transmitted to the crew in a timely fashion. There were some issues that could be addressed if there was sufficient time during noncritical phases of the mission, such as while the vehicle was in orbit.

After the first shuttle was launched, it was thought that some tiles might have been knocked loose off of the thermal protection system on the bottom of the orbiter. The resulting discussions among flight controllers and their support staff resulted in no action being taken, but there was little that could be done anyway. The situation in part reminded me of John Glenn's flight when a faulty sensor indicated that the heat shield was loose which could put the vehicle in jeopardy of being burned up upon re-entry. It was at first decided during that mission that John Glenn didn't need to know these facts if there was nothing to be done to solve the problem. This would not be the only time that the "no news is good news" approach was used for spaceflights when there were no available remedies for possible malfunctions. Every time a possible disaster was averted, it seemed like the issue was forgotten and the underlying problems were never fully addressed.

The thermal protection tiles were always a critical component for vehicle safety upon re-entry, and there was consistent damage to the tiles from the fuel tank insulation materials hitting some parts of the rest of the vehicle during launch, starting with the first shuttle launch. Eventually, the worst possible scenario did happen. Tiles were damaged on liftoff on a critical area of the orbiter wing, resulting in the fatal re-entry disintegration of the *Columbia* on February 1, 2003. I bring this particular case up because I was involved in the original discussions of the thermal protection tiles during the first shuttle mission. At that time, it was determined that the crew was not in danger—most likely. However, there were a lot of unknowns and it had been simulated that missing tiles in critical locations on the wings could cause a "zipper" effect, allowing extreme heat to travel rapidly through the wing and compromise the vehicle. In the case of the *Columbia*, that exact case did happen, and the wing structurally failed. After the *Columbia* tragedy, a method was put into place to investigate suspected tile damage by utilizing cameras that would view the underside of the orbiter while the vehicle was in orbit. A crew spacewalk to repair the tile would be conducted if necessary, and if the damage could not be repaired, the orbiter would rendezvous with the ISS and wait for a rescue mission. This was not a high tech solution, and yet, why NASA didn't employ these methods earlier in the shuttle program is a mystery.

I have the same feeling when I think about my colleague Dick Scobee and the other astronauts who lost their lives in the *Challenger*. Again, a known problem in the solid rocket booster seals was ignored because it didn't result in tragedy yet, until it did....on January 28, 1986. When I worked at NASA, none of my colleagues were knowledgeable in solid rocket booster technology. Morton Thiokol had the expertise necessary to build reliable rockets and determine safe conditions for launch. Warnings were ignored on that cold January morning, and the worst possible result occurred. What hurt the most was that later, it was discovered that the crew compartment was still intact, and the astronauts were at least initially aware of their dilemma for at least some time after the explosion. Several seconds later, the compartment impacted with the ocean, which killed them all. I no longer worked at NASA at that time, but that didn't help the anger and hurt feelings that a disaster might have been avoided. In addition, the crew compartment had no chance of survival, no parachute, and no control mechanism for soft landing in the water. The remainder of the shuttle launches still had no way for the astronauts to survive under similar circumstances in the launch sequence. It was determined that upgrades were too expensive and would add too much weight to the launch vehicle. Luck prevailed, and no other similar launch accidents happened.

My work experiences at NASA demonstrated the positive and negative of the way decisions were made in the shuttle program, which I think was an extension of other programs that came before. Funding and scheduling pressure was a constant in all programs. As it turned out, it was not the failures of some of the more complex or cutting-edge systems that caused these fatal accidents but rather, existing, nagging unsolved problems and the breakdown of human communication and decision-making. Time will tell if these same sorts of issues affect future NASA or private enterprise endeavors.

It is important to focus on the positive effects of these space efforts and understand that focusing on space science and the investigation of celestial bodies is essential for the future of humankind and the preservation and betterment of Earth. Spaceflight is as exciting as it is dangerous, which is why so many people are drawn to it, both in reality and in their love of science fiction.

University of Washington
Tacoma, WA

Linda Dawson

Acknowledgements

I would like to thank Nora Rawn of Springer Publishers for giving me a chance to tell this story. Also, my sister Judi Brodman for her edits and encouragement and my husband Allan for his constant support with this project, even though it meant giving up weekend getaways and vacations.

Contents

About the Author

Linda Dawson received her B.S. in Aerospace Engineering from MIT and an M.S. in Aeronautics and Astronautics from George Washington University at NASA Langley Research Center in addition to completing post-graduate studies in Aerospace Engineering at the University of Washington. She is a Senior Lecturer in Physical Science and Statistics at the University of Washington, Tacoma. Dawson served as Aerodynamics Officer for the NASA Houston Mission Control Center Ascent and Entry Flight Control Teams during the first space shuttle mission (see picture below). During orbital phases, she served as an advisor on the impact of system failures on the orbiter's re-entry trajectory and configuration. From re-entry through touchdown, she was responsible for monitoring the orbiter's stability and control, advising the crew of any necessary corrective actions. Additionally, she serves on the Education Committee and the Space Committee for the Museum of Flight in Seattle, WA.

FDO

1

The New Space Race

Keywords Asteroid Redirect Mission • Chang'e spacecraft • Charles Bolden • Cold War • European Space Agency (ESA) • NASA • NASA Authorization Act • RD-180 • Space Act Agreements (SAA) • Space Exploration Technologies Corporation (SpaceX) • Space Launch System (SLS) • Space race • Space shuttle • Sputnik

> "Space exploration is a force of nature unto itself that no other force in society can rival. Not only does that get people interested in sciences and all the related fields, [but] it transforms the culture into one that values science and technology, and that's the culture that innovates." [1]
>
> –Neil Degrasse Tyson (2012)

There is no longer a definable space race. It ended with the US landing on the Moon and the first human (Neil Armstrong) walking on its surface on July 20, 1969. The space race of the 60s was a clear political and technological race to the Moon between countries representing competing ideologies—democracy in the United States and communism in the Soviet Union. It was an exciting and tense time with political overtones and aggressive posturing that threatened the possibility of nuclear war.

Resources were plentiful in the Moon race due to these external geopolitical pressures ($7–$9 billion over the 5 years following 1961). Given almost unlimited resources in 60s dollars, it became more of a technological race against time. The endgame was succinctly stated by President Kennedy, in a

[1] Tyson, Neil Degrasse. 2012. Space chronicles: why exploring space still matters [audio – radio]. NPR Radio. [Internet] [cited 2016 Mar 17]. Available from: http://www.npr.org/2012/02/27/147351252/space-chronicles-why-exploring-space-still-matters.

L. Dawson, *The Politics and Perils of Space Exploration*,
Springer Praxis Books, DOI 10.1007/978-3-319-38813-7_1

speech to Congress on May 25, 1961: "This nation should commit itself to achieving the goal, before the decade is out, of landing a man on the Moon and returning him safely to Earth."[2] The rest, as they say, is history.

Since the great space race ended, there have been no comparable challenges for the US space program. Progress has continued, but at a slower pace. There have been many achievements and benefits to society from NASA and ESA ventures such as the space shuttle program, the International Space Station, unmanned missions to planets and comets, and the Hubble Telescope's amazing views of the universe, to name just a few. Ambitious new visions are now being posed by commercial endeavors, some in response to prizes offered by NASA. These may not be of the same scale as the 60s space race, but any of these efforts could impact the future of space exploration and contribute to a political advantage in the United States.

The world players in space exploration have changed slightly today, with the traditional superpowers still leading the pack in space efforts—the United States, China, Russia, Japan and the combined European Union. Many national efforts are focused on becoming a regional leader in space technology in order to gain an advantage in Earth science, security and communication. The world situation in some regions is tense, with some countries once again using missiles to demonstrate their military force. In this environment, collaboration among nations is necessary to unite countries in working towards their regional and international goals and prevent any further buildup of aggression in space. Especially as space mining technologies continue to develop, clear operating principles will be needed to prevent strife.

In general, however, momentum is shifting away from national space agencies. The individual faces behind the new space frontier are ambitious businessmen and entrepreneurs. Many grew up as "space cadets," in love with the idea of space travel, launching model rockets in the backyard and thinking that they could build their own rockets someday. With start-up companies and support from NASA, the dream is becoming reality. It will be interesting to watch how this combination of NASA, private enterprise, and other partnerships will combine and create an exciting future for us all.

The future of space exploration is bright. The industry is on the verge of exploring a new frontier—Mars and beyond—with both manned and unmanned missions that will utilize new methods and spacecraft technology developed by a host of new participants. All global citizens will be benefactors of this next phase of Earth's journey into outer space.

---

[2] John F. Kennedy Presidential Library and Museum. [Internet] [cited 2015 June 07]. Available from: http://www.jfklibrary.org/JFK/JFK-in-History/Space-Program.aspx.

The Changing Landscape of Outer Space

Over 50 years ago, outer space was seen as the next frontier for humans to investigate and explore. In addition to other benefits, travel into space would give researchers valuable information about our own planet and what lies beyond. High altitude measurement devices could transmit, for example, more accurate weather data and observational images. Not much was known about the outer space environment and whether humans could even survive travel back and forth from space. It truly was an unknown environment filled only with images from science fiction depictions.

The political landscape after World War II became tense with the development of the Cold War. America initially approached space exploration somewhat cautiously so as not too appear aggressive. The USSR, however, did not have the same strategy and instead pushed forward aggressively, launching Sputnik, the first artificial satellite into orbit. The space race was born, and Americans felt and heard the humiliation every 90 minutes as Sputnik passed overhead, beeping in a foreign language. The next 10–15 years were filled with a series of space firsts, along with each nation's individual successes and failures. The initial rush to explore and dominate outer space was driven more politically than scientifically, and even today, as we move forward into a new age of space development, many of the initiatives are still driven by political motives.

Decades after the space race officially ended, the world today stands at the threshold of a second Space Age and a new type of space race. In the decades after the Moon landing, entire networks of communication and defense satellites were launched, both bringing the world much closer together and placing it under constant surveillance. Humans have lived and worked in outer space for long periods of time, allowing scientists to study long-term effects of experiencing nearly zero gravity aboard space stations. Hundreds of significant experiments have been conducted in laboratories, either in the payload bay of the space shuttle or in space stations, most recently the cooperative effort of the International Space Station. There is now an increased awareness of space debris. So many objects have been placed in orbit around Earth that there is concern about the irreparable damage that satellites or other orbital debris can impact on other spacecraft or stations in orbit. Several countries have participated in both manned and unmanned missions in space, demonstrating a variety of initiatives covering the many facets and applications of outer space exploration.

In the United States, innovative NASA and private sector programs are transforming the space industry. The opportunities for exploration and economic development of the Solar System are expanding. Space technologies have become an integral part of our economy—telecommunications, imaging, and global positioning satellites all formed on the basis of over 50 years of research and development by NASA and other government agencies such as the Department of Defense. Over the next decades, NASA will continue to provide the programs and investments necessary to expand our missions farther from Earth.

NASA's next objectives for exploration include visits to asteroids and Mars, tasks involving more complex technologies and planning than any previous space missions. Successful entrepreneurs worldwide have spent millions of dollars to develop systems that are aimed at exploring and exploiting outer space. More than 50 years after NASA was created, its goal is no longer just to reach a destination in outer space but rather to develop the capabilities that will allow Americans to explore and expand their economic horizons beyond Earth. With the combined talents of government and the private sector, the next journeys beyond Earth will come quicker and will integrate new industries and technologies in the process.

NASA has a legislated responsibility to "encourage, to the maximum extent possible, the fullest commercial use of space."[3] As part of this responsibility, NASA has partnered with private sector individuals and US companies investing in space exploration. In addition to American efforts in space exploration, several countries are now competing for international and regional prestige in the demonstration of space technologies and space science. A new Space Age is well on its way.

The Politics of Space Exploration: NASA and the United States

Space exploration goals and missions are formulated through scientific analysis and agenda-setting by researchers, budget and policy decisions of governments, and through private industry. NASA's funding of colonization or deep space exploration is based on the priorities of the government of the United States in terms of dispensing resources. Each budget item requires explanation and support, and there is also lobbying for the highest level of funding.

The process for budget approval starts with the president submitting an annual budget request to Congress. Leading up to this, all appropriate

[3] NASA.gov. National Aeronautics and Space Act. [Internet] NASA.gov Pub. L. No. 111-314; Dec. 18, 2010 [cited 2016 Mar 22]. Available from: https://www.nasa.gov/offices/ogc/about/space_act1.html

agencies have reviewed their programs and submitted estimates of resources necessary to accomplish their goals. Review committees assemble the requests and discuss priorities with the president, resulting in the final budget request. This process can take an entire year and is a complex process with many levels of review.

The upcoming space budget has been proposed through the NASA Authorization Act for 2016 and 2017, a US law that authorizes NASA's budget with specific line items and policy guidelines. The budget has to pass through Congress for editing before final approval. The proposed budget commits to the development of future efforts such as the Space Launch System (SLS) and Orion. At the same time, "it supports our commitment to once more launching American astronauts, on American rockets, from American soil," states the bill's lead sponsor, Congressman Steven Palazzo (R-Miss.).[4]

The Authorization Act demonstrates strong Congressional support for NASA's success, reiterating the importance of American leadership in space. It aims at regaining national pride in space exploration and enforcing national security through a clear and demonstrated financial plan that supports a clear roadmap for future space efforts. The majority of the funds proposed in the 2016–2017 NASA budget are slated for space exploration and spaceflight technology. The US House of Representative's Committee on Science, Space, and Technology released a statement in April 2015 supporting a more balanced budget that reflects the core mission of NASA with programs in science, aeronautics and spaceflight. It also stated that Mars should be NASA's primary goal.

Chairman Lamar Smith stated "For more than 50 years, the US has led the world in space exploration. We must ensure that the US continues to lead in space for the next 50 years. Astronauts like John Glenn, Neil Armstrong, Buzz Aldrin, Gene Cernan and Sally Ride are household names and national heroes. And today's astronauts inspire American students to study science, technology, engineering, and mathematics and to reach for the stars. Space exploration is an investment we must continue to make in our nation's future."[5]

The bill made its way through the House of Representatives in April 2015 with markups for proposed cuts. NASA Administrator Charles Bolden, an ex-astronaut who has viewed our planet from outer space, responded to the planned budget reductions that he felt would severely impact NASA's Earth science program: "The NASA authorization bill making its way through the House of Representatives guts our Earth science program and threatens to

[4] Committee Plans to Restore Balance to NASA's Budget. US Official News. April 27, 2015.

[5] Committee Plans to Restore Balance to NASA's Budget. US Official News. April 27, 2015.

set back generations' worth of progress in better understanding our changing climate, and our ability to prepare for and respond to earthquakes, droughts, and storm events."

In addition to these cuts, the edited bill underfunded some space technologies thought to be critical to the US leading in the next step space missions, including those leading to Mars.[6]

The Republican-led House Committee on Science, Space, and Technology claims that the cuts in the budget "restores balance" to NASA.[7] The agency was given more than it requested for certain space programs, such as the Space Launch System (SLS) and future missions to Mar and Europa, one of Jupiter's moons. On the negative side, the Earth science budget ended up being cut by almost 40 % from previous levels that had been reached during President Obama's administration. Earth science projects include the study of soil moisture and drought conditions, ocean currents, atmospheric carbon dioxide levels, and climate change indicators such as surveying polar ice.[8] Some think that cuts to these projects are being proposed because of political controversy over climate change. However, it can be argued that the study of Earth science helps us to better understand our own planet and needs to continue at a high rate not only to help this planet but to better understand future destinations and what lies beyond in deep space.

The House committee also voted to cut $240 million from NASA contracts to SpaceX and the Boeing Co., companies that were working to develop new vehicles to shuttle astronauts to the International Space Station (ISS).[9] After the space shuttle program ended in 2011, the only way that American astronauts or supplies could travel to the ISS was on another nation's rocket, most often Russia's. When the relationship between Russia and the United States became tense in recent years, Congress forced the Pentagon to stop buying Russian rocket engines for launching American military and intelligence satellites into space. There was also concern about the United States' sole dependence on Russian transports for our astronauts. Bolden said the first SpaceX or Boeing commercial space flight could be delayed a year to 2018. "If they don't fund it fully, then some milestones won't be done on time and we won't fly in 2017," Bolden said. "And we'll write another check to the

---

[6] Shepherd, Marshall. Cutting NASA's earth science budget is short-sighted and a threat. The Washington Post. 01 May 2015.

[7] Smith, Marcia S. Intense partisanship over NASA resurfaces on house committee. 01 May 2015. [Internet] [cited 2016 Jan 19]. Available from: http://www.spacepolicyonline.com/news/intense-partisanship-over-nasa-resurfaces-on-house-committee

[8] Petersen M. NASA official criticizes cuts to budget in visit to L.A. area. Los Angeles Times. 29 May 2015.

[9] Petersen M. NASA official criticizes cuts to budget in visit to L.A. area. Los Angeles Times. 29 May 2015.

Russians."[10] In June 2015 the Pentagon stated that Russian engines will still be needed for at least a few more years for national security missions. The turnaround on the previous ban has caused controversy in Congress.[11]

The final approved version of the 2016–2017 NASA budget passed before the end of 2015, allotting NASA close to $19.3 billion, funding programs at or above their original request. The budget authorizes full funding for exploration systems that will take Americans to the Moon and Mars as well as the Commercial Crew Development, a program funded by the US government and administered by NASA to encourage private enterprise to develop crew vehicles to travel to low Earth orbit.[12] There will be sufficient resources to keep the new Space Launch System on track with manned missions proposed as early as 2021. US Senator Bill Nelson, Senate Commerce Committee and long-time supporter of deep space exploration, said of the budget: "We are going back into space with Americans on American rockets, and we are going to Mars."[13]

NASA's Goals for 2016–2017

NASA's immediate goals focus on taking major steps to pave the way to Mars. The table shown in Fig. 1.1 displays the space science line items and their associated budgets in millions of dollars for the fiscal year (FY) 2016. Over $5 billion is devoted to furthering space science, looking out to the universe and future destinations as well as looking back toward Earth to further our understanding of our own planet. This fundamental piece of NASA research is a vital component to our knowledge of the universe.

Another nearly $5 billion of the fiscal year (FY) 2016 budget is devoted to space exploration, including the development of vital systems for future programs as well as supporting commercial spaceflight and transportation to the International Space Station (ISS). Space operations to support existing activities primarily in the ISS take up another $4 billion in the budget.

NASA has specific goals in the short term to build on past successes and progress to the next level in the major mission to Mars. Figure 1.2 portrays NASA's projected accomplishments for FY 2016. The Space Launch System

[10] Petersen M. NASA official criticizes cuts to budget in visit to L.A. area. Los Angeles Times. 29 May 2015.

[11] Steven, LM. Pentagon fights ban on Russian rockets. International New York Times. 04 Jun 2015.

[12] Berger, Eric. House budget authorization mark-up slashes $500 million from NASA's Earth science programs. Houston Chronicle. 24 April 2015.

[13] NASA gets boost in Congressional budget deal. Targeted News Service. 2015 Dec 16.

	FY2016
Science	$5,288.6
Earth Science	$1,947.3
Planetary Science	$1,361.2
Astrophysics	$709.1
	$620.0
Heliophysics	$651.0
Aeronautics	$571.4
Space Technology	$724.8
Exploration	$4,505.9
Exploration Systems Development	$2,862.9
Commercial Spaceflight	$1,243.8
Exploration Research and Development	$399.2
Space Operations	$4,003.7
International Space Station	$3,105.6
Space and Flight Support (SFS)	$898.1

Fig. 1.1 NASA space science budget allocations for FY2016 (Image from NASA.gov. Fiscal Year 2016 Budget Estimates. [Internet] NASA.gov Budget Summary; [cited 2016 Apr 18]. Available from: http://www.nasa.gov/sites/default/files/files/NASA_FY2016_Summary_Briefing.pdf)

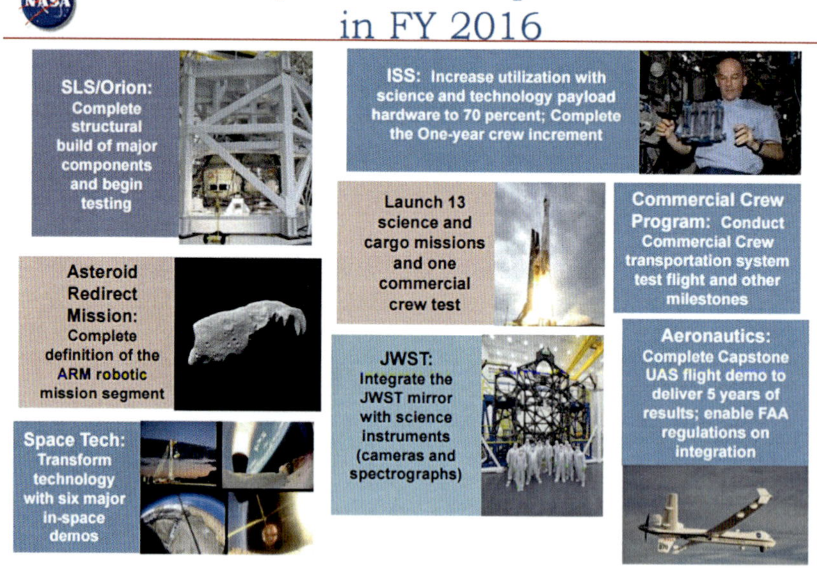

Fig. 1.2 Anticipated accomplishments by NASA in FY2016 (Image from NASA.gov. Fiscal Year 2016 Budget Estimates. [Internet] NASA.gov Budget Summary; [cited 2016 Apr 18]. Available from: http://www.nasa.gov/sites/default/files/files/NASA_FY2016_Summary_Briefing.pdf)

(SLS)/Orion vehicle and its supporting systems provide the rocket transport and crew vehicle for Mars missions. The Asteroid Redirect Mission (ARM) is a component of the overarching Mars mission. The connection between grabbing an asteroid and going to Mars has eluded Congress and the space community. However, two aspects of the mission are popular—finding and tracking asteroids, and the development of high power solar electric propulsion. In addition, NASA also advertises the ARM as a future application of altering the path of an asteroid threatening Earth by a very small amount, by using the gravitational pull between the asteroid and a nearby spacecraft. This is called a "gravity tractor."

Other important milestones include the Commercial Crew Program, which will conduct test flights and other milestones to provide safe and affordable commercial transportation launched from American to and from the ISS to deliver crew and supplies and provide for additional scientific research in the orbiting laboratory.[14]

The Politics of International Space Exploration

As stated earlier, space exploration goals and objectives are formulated as a result of scientific analysis and the decisions of governments. Many countries are interested in exploring space and being able to colonize and capitalize on resources in deep space, including those found on planets and asteroids. Being first to an area of interest in outer space is a way of gaining political and military advantage. This can be of concern in a time where we are guided by a space treaty signed in 1967. The Treaty on Principles Governing the Activities of States in the Exploration and Use of Outer Space, Including the Moon and Other Celestial Bodies (see U.S. Department of State website for the complete text http://www.state.gov/t/isn/5181.htm) essentially states that outer space belongs to all nations on Earth and should be explored for the benefit of mankind in a cooperative manner. In signing this treaty, the United States and the Soviet Union effectively agreed not to place any nuclear weapons or weapons of mass destruction on a celestial body or orbiting Earth.

In addition, the permanent or temporary establishment of military bases, the testing of weapons, and the holding of military maneuvers on celestial bodies (admittedly a long-shot at the moment, anyway) is strictly forbidden.[15]

[14] NASA.gov. Fiscal Year 2016 Budget Estimates. [Internet] NASA.gov Budget Summary; [cited 2016 Apr 18]. Available from: http://www.nasa.gov/sites/default/files/files/NASA_FY2016_Summary_Briefing.pdf

[15] U.S. Department of State. Treaty on principles governing the activities of states in the exploration and use of outer space, including the moon and other celestial bodies. Bureau of Arms Control, Verification, and Compliance. Signed 27 Jan 1967. [Internet] [cited 2015 June 04]. Available from: http://www.state.gov/t/isn/5181.htm

The United States and about 90 other countries signed the treaty. Almost 50 years later, the treaty is still in force but needs updating to protect against the militarization of outer space. Many countries do not respect borders and agreements on Earth, and a military presence has now expanded into low Earth orbit and beyond. It is well known that China has demonstrated their military capability in Earth orbit, testing laser weapons that disarm satellites with potentially crippling results. The treaty can now be interpreted as allowing certain activities. Now may be the time to be more specific about prohibiting space weapons. China and Russia have been amenable to discussions addressing updating the treaty. However, the United States has stated, as recently as 2006, that the Outer Space Treaty is sufficient in its intent. This is one conversation that is worthy of following as many nations compete for outer space resources.

Space exploration is an expensive undertaking, and the actual number of nations that have made significant steps in pursuing scientific research in space or towards a particular goal are few. We know that China has planned a permanent presence in space and has a very active unmanned and manned space program. Russia still has a stronghold on launch systems and hopes to restore its manned exploration program. It has also indicated an interest in building its own space station. India and Iran have indicated interest in manned space programs.[16]

It is believed that by 2015, as many as 70 countries had participated in some activity associated with space exploration. To be clear, most of the countries were invested minimally, in order to keep up with other countries, particularly in their regional and trade areas. Figure 1.3 is a visual display of countries with government space programs and their financial investment, comparing 2003–2013. Space efforts even on a small level are becoming a standard part of many governmental efforts to invest in the future.

Henry Hertzfeld, research professor at George Washington University's Space Policy Institute, states, "There are a lot of new starts." He does say that it is important to put them in perspective, though. "There are different launch vehicles and different capabilities, too. Comparing a manned capability that India might want to spend some money on with Iran launching a very small, very low Earth orbit satellite is really apples and oranges."[17]

Many countries have also been able to participate in some aspect of space research by having crew members from their country travel to the ISS and carry out work there. Figure 1.4 displays a map showing how many countries have visited the ISS, demonstrating an interest in space travel and space science.

---

[16] Ball, Philip. Time to rethink the outer space treaty. 04 Oct 2007. doi:10.1038/news.2007.142.

[17] Belfiore, Michael. Popularmechanics.com. 30 Sep 2009. [Internet] [cited 2016 Mar 25]. Available from: http://www.popularmechanics.com/space/a12531/4307281/

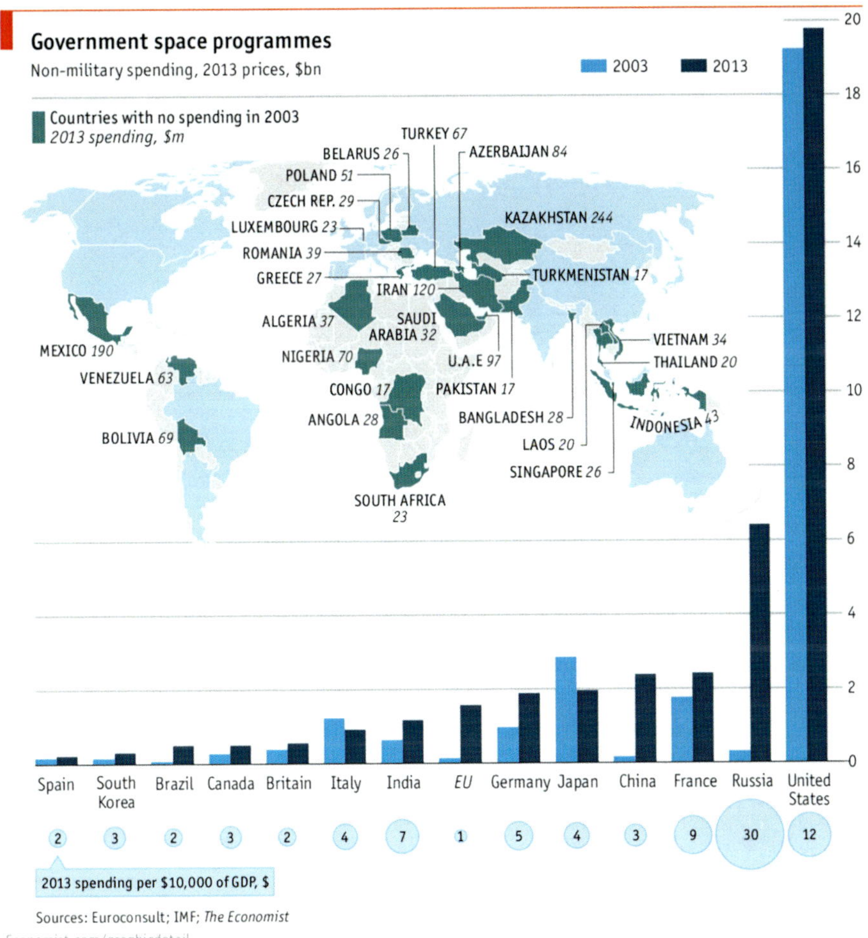

Fig. 1.3 International government space programs (2003–2013) (Image courtesy of www.economist.com)

There is a baseline level of communication and security needs for most developed countries. Satellite networks have become commonplace and a fundamental requirement to connect to the rest of the world. Other countries have the resources and interest to expand beyond the basic level and participate in developing technologies and experiments to further scientific knowledge of outer space. To achieve some worldwide agreement addressing common goals in space endeavors, the International Space Exploration Coordination Group (ISECG) was created in 2007 after 14 space agencies developed a global exploration strategy that presented a shared vision of coordinated efforts for both human and robotic missions to explore the Solar

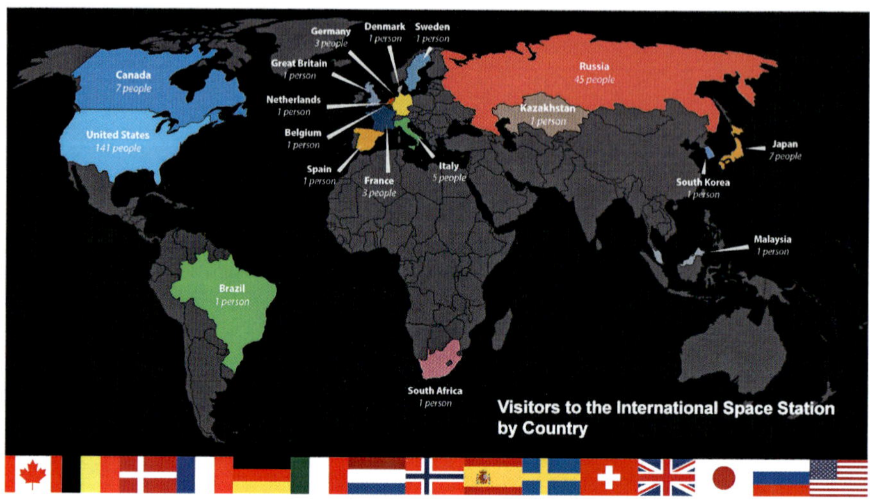

Fig. 1.4 Visitors to the ISS by country (Image courtesy of NASA)

System. Current members of the ISECG are Italy, France, China, Canada, Australia, Germany, the European Space Agency, India, Japan, the Republic of Korea, the United States, the Ukraine, Russia and the United Kingdom. The organization created a Global Exploration Roadmap in 2013 that identifies common goals and objectives as well as a strategy for long-range human exploration (Fig. 1.5).

The advantage to an international roadmap to outer space is to bring consensus information to interested parties in order to keep everyone informed about current and near-future missions and to provide pathways for partnerships. Despite the differences of the involved countries and space agencies, the commonalities are straightforward and consistent, reflecting similar ideas that involve the betterment of humankind. Some of the common approaches are Earth related—enhancing Earth safety and engaging the public in space exploration. Other goals focus on travel to outer space—extending the human presence beyond low-Earth orbit, doing research to enable humans to work and live in space environments, searching for life, and stimulating economic expansion into space that can benefit us back on Earth.[18]

Here is a look at the capabilities of the top and most talked about space-faring nations in what may be a new world order. The race is on for space dominance.

[18] NASA.gov. ISECG. The global exploration roadmap. [Internet]; Aug 2013 [cited 2016 Mar 25]. Available from: https://www.nasa.gov/sites/default/files/files/GER-2013_Small.pdf

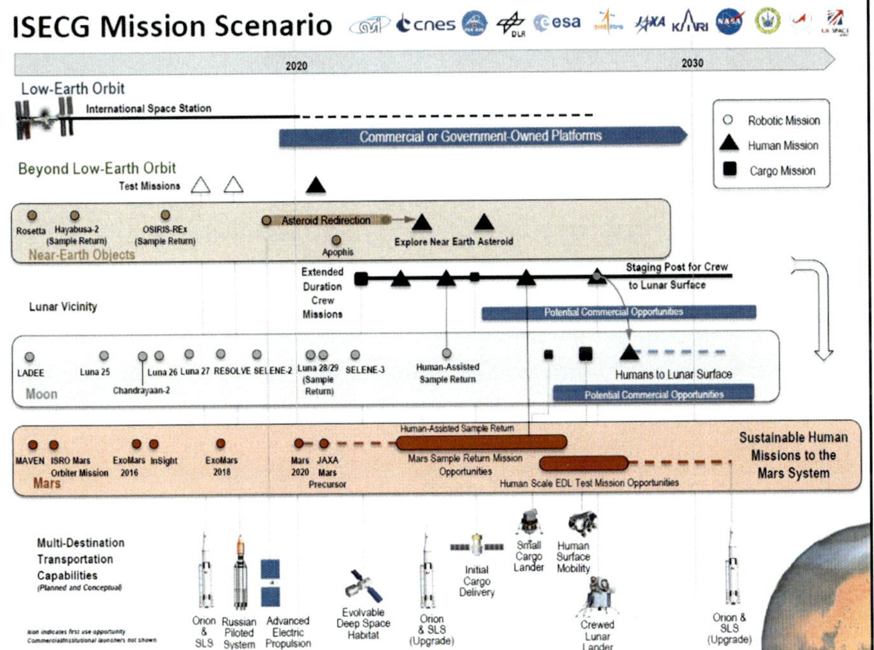

Fig. 1.5 ISECG mission scenario (Image courtesy of NASA)

Russia

Russia is still very much involved in space transport and the scientific exploration of deep space. The Russian Federal Space Agency, Roscomos, is significantly involved in the day to day running of the ISS, shuttling cargo and astronauts back and forth to Earth on Russian spacecraft, as well as providing resources and guidance for ISS operations. As previously discussed, this has become a source of controversy in the United States, particularly since relations between the two countries have become tense. The United States is moving to become once again independent of the Russian space transport services through its Commercial Crew Transport program. However, it was recently revealed how complicated and intertwined the relationship has become between the two space programs.

Russia has always excelled in rocket design and development, and their designs have been reliable and powerful in a variety of missions over the past several decades. Their launch systems are currently used regularly to travel to the ISS, carrying both US and Russian cargo. The RD-180 and the NK-33 rocket engines made in Russia have been part of the US–Russian partnership to transport cargo to the ISS. The RD-180 rockets are powerful and more efficiently designed that any American rocket engines.[19] Historically, Russian rocket design had the advantage of some of the best German rocket scientist minds, who went to work in Russia at the end of World War II, when the German rocket scientists were split between the United States and Russia. The United States was fortunate to receive some of the best guidance and navigation system engineers, including the great Wernher Von Braun, who many feel contributed to the success of landing on the Moon and ending the space race.

The American corporation United Launch Alliance (ULA), the sole supplier of US military rocket launchers, has been purchasing modified Russian RD-180 engines to power the Atlas rockets. The Atlas rockets conduct military launches for the United States, putting spy satellites into space. The United States originally had two rockets for these applications—the Atlas and the Delta. The Atlas 5 is more attractive in terms of performance and price because of the addition of the RD-180 engines.

By the late 1990s, the United States had negotiated a long-term contract with a license to build the RD-180s. By the end of 2014, due to a complicated political situation, Russia banned the use of its rockets for military purposes, allowing only civilian payloads. This complicates the intertwined space partnership between the two countries. The license to build the RD-180s, without less expensive Russian parts, requires a production line that would probably take 5–7 years to set up and a considerable amount of money (an additional $5 billion) to support it.[20] In addition, satellite launches could be delayed more than 3 years (Fig. 1.6).

The other Russian rocket engine of note, the NK-33, was first built in the 1960s to take Russians to the Moon. After the space race was over, some of these engines were mothballed in warehouses. In a similar deal as the RD-180s, dozens of NK-33s were sold in the 1990s to Aerojet General for a little over $1 million. The company also acquired a license to build the engines. The engine was used in the heavy lift Antares rocket, an expendable launch system developed by Orbital Sciences Corporation to launch spacecraft to low Earth orbit and the ISS. Antares was launched for the first time in 2013 and had

[19] Staff, Washington Times. Rocket contracted by NASA explodes after Virginia launch. 30 Sep 2009. [Internet] [cited 2016 Mar 25]. Available from: http://www.washingtontimes.com/news/2014/oct/28/rocket-contracted-by-nasa-explodes-after-virginia-/?page=all

[20] Magnuson, Stew. Costs, benefits of RD-180 rocket engine replacement program debated. National Defense Magazine. Jul 2014. [Internet] [cited 2016 Mar 25]. Available from: http://www.nationaldefensemagazine.org/archive/2014/July/pages/Costs,BenefitsofRD-180RocketEngineReplacementProgramDebated.aspx

Fig. 1.6 Atlas V rocket raised with RD-180 engines (Image courtesy of NASA)

four successful launches until the fifth launch in October 2014, when the rocket failed catastrophically, destroying the vehicle and its payload.

The explosion reinforces the idea that the United States cannot compete with the Russians in the development of reliable and powerful rocket engines and integration into a launch system. Time will tell how this explosion will affect American space exploration plans.[21] The accident investigation report identified an explosion in the liquid oxygen turbo pump in one of the two Antares engines. A definitive reason for the failure could not be determined but was thought to be a result of one of three causes or a combination of any of the following—"inadequate design robustness" of the engine, making it more vulnerable to oxygen fires; foreign object debris inside the engine; or a manufacturing defect with the engine found in previous engine testing. The engines, even though they were old, did not have historical failure data that could have provided valuable information. In addition, there were some issues

[21] Simha, Rakesh K. Russian rocket science eludes US space programme. Russia & India Report. 22 Nov 2014.

regarding risk factors and how they would be communicated and evaluated with the Commercial Resupply Services contract between NASA and Orbital. This last issue is reminiscent of communications failures associated with the *Apollo 1* fire and the space shuttle explosions.[22]

Orbital has now replaced the faulty engine with the RD-181 engine, considered to the best option in terms of availability, performance, and cost. Four RD-181 engines were purchased through a $1 billion contract agreement in 2014 from Energomash, Russia's rocket engine manufacturing company. Many US government officials wanted to block the sale, citing existing sanctions on Russia. However, Washington removed the engines from the blocked list because they proved to be indispensable for US space flights. There were no currently American-made equivalents, and to produce an engine similar to the RD-181 would take at least 5 years.[23] The engines completed a 30-second test firing on a launch pad on Virginia's Eastern Shore at the end of May 2016. The successful test is key to the Antares rocket returning to flight in mid-October 2016, resuming cargo service to the International Space Station (Fig. 1.6).

It is currently thought that the Russian space industry, as well as other government-supported Russian programs, is having financial difficulties. The building of a new launch facility has been delayed citing setbacks due to funding problems and possible corruption. The Federal Space Agency's budget has been cut by 35 % for the next decade.[24]

Igor Komarov, the head of Roscosmos, said at a news conference in April 2015: "The cost of the program's projects has undergone significant changes over the last year given the prevailing economic conditions, changes in exchange rates and changes in the level of inflation."[25]

Other Russian plans include the development of a space station that improves on ISS technology. Roscosmos Chief Oleg Ostapenko says: "We are considering the possible construction of a high-latitude station from which

---

[22] NASA Summary of the ORB-3 Accident Report. NASA.gov. 09 Oct 2015. [Internet] [cited 2016 Jan 16]. Available from:

http://www.nasa.gov/sites/default/files/atoms/files/orb3_irt_execsumm_0.pdf

[23] Staff, Space Daily. US to buy eight Russian rd-181 rocket engines. 15 Mar 2016. [Internet] [cited 2016 June 09]. Available from: http://www.spacedaily.com/reports/US_to_Buy_Eight_Russian_RD_181_Rocket_Engines_999.html

[24] Kottasova, Ivana. Economic crisis at heart of Russia's pride: its space program. 27 April 2015. [Internet] [cited 2016 Jan 16]. Available from: http://money.cnn.com/2015/04/27/news/economy/russia-space-crisis-cosmodrome/

[25] Kottasova, Ivana. Economic crisis hits at heart of Russia's pride: its space program. CNN Money (London). 27 April 2015. [Internet] [cited 2015 June 15]. Available from: http://money.cnn.com/2015/04/27/news/economy/russia-space-crisis-cosmodrome/

90 % of the Russian territory will be visible. It may become a base for prospective lunar expeditions."[26]

Roscosmos plans to spend three quarters of a billion dollars on a new manned spacecraft, which is more than three times less than NASA allotted to SpaceX for the commercially built Dragon space vehicle. Russia plans to launch their new space vehicle in 2021 and use the vehicle to transport crews and supplies to the ISS.[27]

Planned missions for the establishment of a lunar base and an expedition to Mars are well into the future, 20–30 years. The progress depends on a consistent level of funding.

China

Compared to the United States and Russia, China's space program has been slow to develop but is now in the midst of what many view as a competitive space race in Asia. China's first human spaceflight took place in 2003, with China becoming the third nation after the United States and Russia to do so. The rise of the Asian economy has resulted in increased resources for space programs and military efforts. The primary reason for success in this arena is seen to be a desire for dominance in that geographical region, bringing prestige to that nation and encouraging the growth of science and technology.

China has launched multiple vehicles to the Moon in the past decade. A series of Chang'e spacecraft (1–3) were sent to the Moon to orbit and to explore the surface, including a robotic rover landing successfully by the end of 2013. The *Chang'e 4* and *5* spacecraft will be sent to explore the dark side of the Moon. It is expected that a mission to land on the dark side would be attempted by 2020, a feat that no other nation has attempted so far. There is concern that China's progress will translate into dominance over resources on the Moon, including helium 3, a gas than could be used to provide nuclear power without radioactive waste, as well as an abundance of rare minerals and precious metals that are extremely high in price that can be used for a variety of electronics and industrial applications (Fig. 1.7).[28]

China is also interested in developing a space station, a heavy-lift booster rocket, and a new launch site. These can all be seen as peaceful means of space

[26] Roscosmos: High-latitude orbital station may become lunar expeditions' base. Interfax : Russia & CIS General Newswire. 16 Dec 2014.

[27] Russian News Agency. Russia's new manned spacecraft to be 3.5 times cheaper than US Dragon. 22 Jan 2016. [Internet] [cited 2015 June 16]. Available from: http://tass.ru/en/science/851562

[28] Shen L, Hunt K. China to explore 'dark side' of the moon. CNN Wire Service. 21 May 2015.

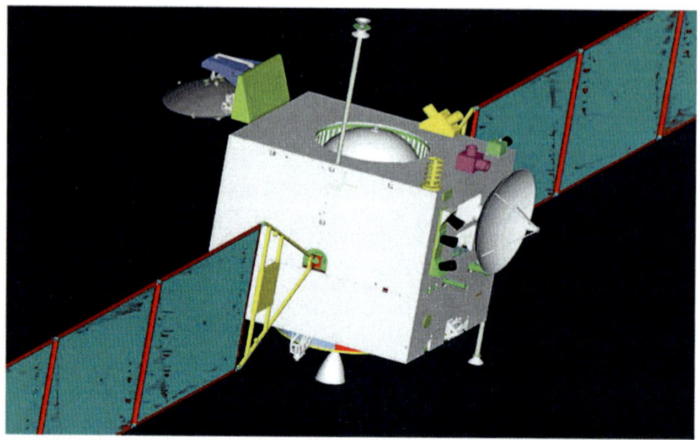

Fig. 1.7 The *Chang'e 1* spacecraft (Image courtesy of NASA)

exploration. However, China's anti-satellite program tested in 2007 was seen as a military threat to the United States and a possible violation of the space treaty. The satellite system is part of the national security for not only the United States but also other nations, including Europe and Russia. China's demonstrated ability to destroy a satellite in orbit by using an interceptor raised not only security concerns but also possible destruction of property caused by the space debris generated from this type of explosion. In May 2015, China also tested part of a new anti-satellite ballistic missile system. These aggressive military actions are part of a larger military posturing by China, and many nations are watching as these events unfold.[29]

The sci-fi movie hit *The Martian* portrays the Chinese space agency and one of its rockets saving the day to rescue astronaut Watney, who is left stranded on Mars. Many wondered if this type of collaboration could occur between nations that are at generally opposed to each other on many issues. In fact, Congress passed a spending bill in 2011 that forbids NASA from working with China, including coming aboard or being involved in the International Space Station, citing a high risk of espionage. It doesn't appear that the relationship between the nations has changed much since then.[30] However, NASA could gain financial support and technological advances by working with the Chinese. Russia is supporting the ISS for a limited timeframe, and it could be of benefit to build the next space station collaboratively with those nations not involved in the ISS.

[29] Asia's space plans worry. Eastern Eye. 2013 Dec 20; Sect. 17.

[30] Dickerson, Kelly. Here's why NASA won't work with China explore space. Tech Insider. 2015 Oct 19.

India

India is intently watching China's progress in space. It has opposed the weaponization of outer space but understands the importance of space exploration to gain political power and to advance science and technology. It has upgraded its space science program in order to gain more prestige in this area.

Some of India's accomplishments include:

- Launching its first military satellite in 2013.
- The Mangalyaan spacecraft successfully orbiting Mars in September 2014 (India becomes the first Asian country to reach the Red Planet).
- Lunar orbiters *Chandrayaan 1* (2008) helping to confirm the existence of water on the Moon. *Chandrayaan 2* to be launched in 2017 will test new technologies and conduct science experiments.
- Future missions to study the Sun's corona (*Aditya-1*) and future plans for human spaceflight.[31]
- Chinese President Xi Jingping and India Prime Minister Narendra Modi recently signing a pledge to explore peaceful cooperation in space. This is a positive step in overcoming decades of mistrust and leading the two countries to a place of mutual respect.[32]

India has managed to accomplish many goals in space on limited resources. Their society has benefited from space research and advanced technology: "Ten or fifteen years back, the loss of life due to cyclones used to be thousands. Today it is one or two."[33] Clearly, spinoffs from technology designed for space travel has made life better for Indians today. India's space program continues to grow, with the launch of larger satellites and a heavy-lift rocket and improved Earth-observation capabilities. There are currently at least ten missions scheduled for launch in 2016, including a seven satellite system aimed at improving road and marine navigation.[34] The future seems bright for India's space program.

---

[31] Working on manned space mission: ISRO chairman. Indian Express. 04 Jan 2015.

[32] Menon, Jay. India's second moon mission to be fully homegrown. Aerospace Daily & Defense Report. 20 April 2015.

[33] Petrou, Michael. The underdog explorers. Maclean's. 31 Aug 2015. Maclean's. V.128 p. 65.

[34] Menon, Jay. India plans heavy launch schedule in 2016. Aviation Week & Space Technology. 01 Jan 2016.

Fig. 1.8 Canadarm2 removing unpressurized payload from Japan's HTV-2 Transfer Vehicle (Image courtesy of NASA)

Japan

Japan reacted to China's anti-satellite system by ending their previous ban on military activities in space (2008). Tokyo has increased its space efforts with a new launch vehicle and renewed efforts in space science and human spaceflight. This includes scientific research on its Kibo module on the International Space Station. Japan's H-II Transfer Vehicle spacecraft now provides the only non-US and non-Russian transport vehicle able to ferry supplies to the ISS (Fig. 1.8).[35]

Japan also fears that China's space accomplishments might affect its own technological dominance. One of Japan's more ambitious space missions is the Hayabusa 2, launched in December 2014. The purpose is to put four landers on an asteroid by 2018 and return soil samples to Earth. Japan's lunar missions (Kayuga) explored the Moon's surface—its geography and composition.[36]

[35] Moltz, James C. It's on: Asia's new space race: while NASA and the European Space Agency gets most of the world's attention, China, Japan and India are racing for the heavens. The Daily Beast [New York]. 17 Jan 2015.

[36] Howell, Elizabeth. Japan Aerospace Exploration Agency: facts and information. 05 Sep 2013. [Internet] [cited 2015 June 16]. Available from: http://www.space.com/22672-japan-aerospace-exploration-agency.html

Other Asian Space Activities

Other Asian countries have also shown enhanced space activity. North Korea successfully launched a primitive satellite from its *Unha 3* rocket in December 2012. South Korea responded with a launch of a more sophisticated satellite in January 2013, using a Russian first stage on the rocket. South Korea has also announced accelerated plans for a larger launcher and a lunar research mission in the 2020 time frame. Several smaller efforts are being made in Singapore (satellite manufacturing and communications) and Vietnam (satellite development and technology development with Japan). [37]

The European Space Agency

The European Space Agency (ESA) coordinates the civilian space activities for 20 member states, with centers scattered in several European countries. ESA is a major contributor to the ISS and flies astronauts to the orbiting station on a regular basis. A cargo vehicle, called the Automated Transfer Vehicle (ATV), ferries supplies to the ISS as well. ESA spacecraft have explored several planets in the Solar System, including Saturn and Mars. A number of its satellite programs explore Earth and other celestial bodies—notably, Mars Express, which is in orbit around Mars, and the Rosetta spacecraft, which took images of several asteroids on its way to its destination, Comet 67-P. Finally, ESA is working on the next generation GPS technology.

ESA has its own rockets—the most famous being the Ariane, which has gone through several iterations since it was first launched in December 1979. The rocket is an example of a reliable and affordable launch system that is financed through commercial exploits.[38]

[37] Moltz, James C. It's on: Asia's new space race: while NASA and the European Space Agency gets most of the world's attention, China, Japan and India are racing for the heavens. The Daily Beast [New York]. 17 Jan 2015.

[38] Howell, Elizabeth. Howell, Elizabeth. European Space Agency: facts and information. 27 Aug 2013. [Internet] [cited 2015 June 16]. Available from: http://www.space.com/22562-european-space-agency.html

Other Space Agencies

Australia defined a space policy in 2013, expressing their interest in cooperating with the United States in agreements regarding space awareness, tracking of orbital debris and collision prevention in space. Other countries with interest in space activities are:

The Israel Space Agency

This was established in 1983, has funds devoted to the Venus Project and other science and technology development for space interests.

Iran

Launching satellites and other spaceflights since 2005, Iran has had joint projects with Russia, Thailand and China, launching satellites designed for research and communications.

The United Kingdom

This country created its own space agency in 2010. The UK paired up with the European Space Agency, where they have been involved in scientific research and the launch of low Earth orbit rockets.[39] Spaceport planning is a main area of interest for the UK space sector.

The United Arab Emirates also has a very newly formed space agency, as of only 2014, yet is eagerly expanding into the field and taking the space sector very seriously as an engine of economic growth.

The Politics of the New Space Age

International politics today is very different than politics in the 1950s and 1960s. The conflicts that we face today are multi-faceted, with threats that are not necessarily defined by geographic location. In the past, space achievements were restricted to the United States and the Soviet Union and were

[39] Ten leading space programs around the world. CBC News. 04 Nov 2013. [Internet] [cited 2015 June 16]. Available from: http://www.cbc.ca/news2/interactives/space-programs/

a sign of their political power and military prestige. As time went on, more countries became interested in space exploration not only to demonstrate prestige but to discover more about the universe and to provide the technology for better communication, navigation, and business opportunities for their countries. Several countries began to cooperate on space missions, and ultimately the International Space Station came into being as a testimony of this international effort. Fear of countries staging war either in space or from space have not come to fruition, despite the occasional demonstration of such capabilities. Regional demonstrations of space dominance seem more likely, however, none of these qualify as "races." It can be said that countries with power and a natural competition with each other consider space as another platform to demonstrate excellence with advantages that are not necessarily focused on the military.

China's recent aggressive posturing with laser weapons, anti-satellite missiles and jammers are of concern to the United States as well as other countries. Their plans for high altitude hypersonic vehicles, as well as low-Earth orbit platforms, create a military scenario where China desires to control the airspace closest to Earth. The United States has to match these capabilities along with having the ability to detect China's use or testing of their high-tech "weapons." China's activities border on the aggressive use of weapons in airspace defined for peaceful purposes in the space treaty. At this time, the United States is monitoring and creating a strategic space plan to deal with perceived threats. At the same time, China's neighbors are watching their progress with similar interest, if perhaps with even greater alarm. Certainly the country has shown an aggressive approach to naval expansion, and it could be that Sino space activities may follow the same model. Claiming territory—and any associated potential mineral deposits—will surely be tempting.

US dependence on Russia's transportation to the International Space Station is of concern as well. However, as far as space efforts go, Russia has been nothing but cooperative and thus far has provided no major reasons for additional concern. Time will tell how this relationship will play out in outer space, since it is increasingly volatile down on Earth.

Overall, the international level of effort of space exploration today is remarkable. By all levels of measurements, it seems that the status of both interest and level of development and planning for space travel both in the United States and around the world is very high. The future is bright for space research and exploration. The primary thing that holds countries back from pursuing further efforts is their financial status; however, the desire to know more and reap the benefits of technology in space for all nations seems universal.

The more interesting and dynamic space race is among the new faces of space entrepreneurs. Elon Musk's Space Exploration Technologies (SpaceX) is now a low-cost provider of low-Earth orbit transport, and Jeff Bezos's Blue Origin is on the verge of establishing his company as a major contender in space exploration. Other companies are developing new technologies and ideas to make space travel easier and more affordable. These innovators will define the future of space exploration. Instead of a geopolitical struggle, the new space race is a test between private-sector engineers and commercial interests, with the resulting competition having the potential to drive forward the needed innovations for manned missions to succeed, while international collaboration by and large remains the watchword in government-funded space missions driven by the desires of the global scientific community of researchers.

The space frontier will be explored with enthusiasm and gusto in the next decade, with humankind being the ultimate benefactor. The goal of landing humans on Mars is getting closer to reality.

2

The Commercial Space Race

Keywords Blue Origin • Boeing Co • Commercial Crew Development (CCDev) • Commercial Crew Program (CCP) • Dream Chaser • Elon Musk • Google Lunar XPrize • Orbital ATK • Paragon Space Development Corp • Sierra Nevada Corporation (SNC) • SpaceX Dragon • SpaceX Falcon • United Launch Alliance (ULA) • XPrize

> "Since Yuri Gagarin's and Al Shepard's epoch flights in 1961, all space missions have been flown only under large, expensive government efforts. By contrast, our program involves a few dedicated individuals who are focused entirely on making spaceflight affordable."[1]
>
> –Burt Rutan (2004)

Decades ago the thought of a private company launching rockets with both manned and unmanned payloads was unthinkable. Today, dozens of companies are developing space vehicles and their associated technologies. Creativity abounds in thinking of new ways of performing required tasks in space. Without restrictive guidelines that sometimes occur with government contracts, these companies are feeling a new freedom in their initially self-funded efforts to create methods that might prove to be easier, more efficient, while using fewer resources. The American space program will benefit from all of these creative efforts.

National funding for the US space program remains tied to the goals of government funding for each administration as well as the power of political

[1] David, Leonard. Private spacecraft blast offs June 21. 02 June 2001. [Internet] [cited 2016 Apr 26]. Available from: http://www.cnn.com/2004/TECH/space/06/02/private.space/

L. Dawson, *The Politics and Perils of Space Exploration*,
Springer Praxis Books, DOI 10.1007/978-3-319-38813-7_2

lobbyists. In view of a lag in funding, interest was generated in the private sector through prize money along with a renewed public interest and support for space-related activities. When the shift began to private enterprise, a sector of the scientific community was concerned that the new space ventures could be reckless and might lack the checks and balances that NASA had put in place. In addition, barriers to space enterprise are in place in the form of government regulations that can bog down some efforts, increasing expenses and delaying timelines.

So far, the private ventures have brought an incredible variety of new and exciting participants into solving the challenges of our next goals in space.

The Commercial Development of Outer Space

For decades, NASA has led the American space program. As a government agency, NASA funding has been dependent on government priorities, and over time has experienced periods of increased and reduced support depending on the state of the economy. Over time, this has led to reduced goals in space exploration and extended timelines. The breaking point came when it became obvious that there would be no American vehicle to transport American astronauts to the International Space Station once the space shuttle was retired. In addition, private companies were starting to think of outer space as a potential commercial opportunity. Some commercial efforts responded to the need for rockets and vehicles to transport cargo and crews to the ISS and deliver satellites into orbit. Other companies had new visions to provide a space tourism industry or create new technologies to make space transport easier or to explore new options such as mining asteroids.

Some of these companies are highlighted in the following sections. In addition, the history of increased private engagement is chronicled in this chapter.

The Space Exploration Technologies Corporation (SpaceX)

One of the most successful entrepreneurs today is Elon Musk, multi-billionaire and currently CEO of SpaceX (Hawthorne, CA) as well as Tesla Motors, and chairman of SolarCity. He founded SpaceX in 2002 in order to design, manufacture, and launch rockets and spacecraft. The ultimate goal of the company is to enable humans to live on other planets. SpaceX has grown from only 160 employees in 2005 to currently over 3500 employees.

Musk says that he originally founded SpaceX in order to support NASA in building a greenhouse on Mars. However, since there are no immediate plans

Fig. 2.1 ISS-31 SpaceX Dragon commercial cargo craft is grappled by the Canadarm2 (Image courtesy of NASA)

to go to Mars, he decided to start his own space company. His goal was to improve rocket technology and develop his own launch system. His persistence, vision, and long-term success formed the basis of a collaborative effort forged with NASA and through lucrative government contracts.

SpaceX has developed the *Falcon 1* and *Falcon 9* launch vehicles, with the goal to eventually have those vehicles become completely reusable. They have also designed and built the Dragon spacecraft, which is flown into orbit by the *Falcon 9* launch vehicle to supply cargo to the ISS (Fig. 2.1). Musk explained that his business strategy is to transport cargo to low Earth orbit, saving time and money. He estimates that the cost per pound of payload is about $1000, whereas in the past, it was $5000–$10,000 per pound.

In 2012, SpaceX's *Falcon 9* launched the Dragon space capsule as payload to the ISS, making history as the first private company to dock with the ISS and deliver cargo. Since then, the company has made several flights to the ISS for NASA. The company is also interested in the development of reusable rockets. Rockets generally only have a one-time use, falling off into the ocean as trash. The space shuttle was an exception, with the solid boosters being retrieved and refilled for subsequent flights. Musk's objective was to add landing legs to the rockets so that they can land on a barge and be reused. "We have to do something dramatic to reduce the cost of getting to space,"

Musk has said. "If we can get the cost low, we can extend life to another planet. I want to help make humanity a spacefaring civilization." [2]

SpaceX's milestones include:

- The first privately funded liquid-propellant rocket, the *Falcon 1*, reached orbit (September 28, 2008).
- The first privately funded spacecraft, the Dragon, successfully launched, orbited, and was recovered (December 9, 2010).
- The first private company to send a spacecraft, the Dragon, to the ISS (May 25, 2012).
- The first SpaceX satellite delivery into geosynchronous orbit (December 3, 2013). [3]
- The first landing back on Earth of a first stage of a rocket after launching and sending a payload into orbit (December 21, 2015). After this event, Elon Musk told reporters: "I do think it's a revolutionary moment. No one has ever brought an orbital class booster back intact. We achieved recovery of the rocket in a mission that also deployed 11 satellites. This is a fundamental step change compared to any other rocket that's ever flown."[4]

The road to space for SpaceX has not been without setbacks and failures. There have been numerous launch aborts as well as unsuccessful landings of the Falcon launch systems. On June 28, 2015, a SpaceX *Falcon 9* rocket exploded on launch, destroying an unmanned Dragon spacecraft carrying food, care packages as well as scientific and electronic equipment and resources for the ISS. The reason for the explosion was a faulty mechanical strut, demonstrating the importance of quality control and multiple backup systems for each piece of a launch system. About 6 months later, in December 2015, the *Falcon 9* rocket returned to flight with a successful mission delivering 11 satellites to orbit for their customer Orbcomm. This time, the first stage booster returned to the launch site successfully, rather than to a barge in the ocean.[5]

---

[2] Lee, Rhodi. SpaceX milestones: how Elon Musk brought a company to the forefront of spaceflight. Tech Times. 26 May 2015.

[3] Lee, Rhodi. SpaceX milestones: how Elon Musk brought a company to the forefront of spaceflight. Tech Times. 26 May 2015.

[4] Wall, Mike. Wow! SpaceX lands orbital rocket successfully in historic first. 21 Dec 2015. [Internet] [cited 2016 Jan 21]. Available from: http://www.space.com/31420-spacex-rocket-landing-success.html

[5] SpaceX conducts return-to-flight launch, rocket lands on ground. Cihan News Agency. 22 Dec 2015.

In September 2016, the Falcon 9 exploded again, this time during a prelaunch test. Boeing-Lockheed has raised concerns to the Pentagon about "going with the lowest bidder" on sensitive national security launch contracts. Time will tell how this will be sorted out.[6]

Sierra Nevada Corporation (SNC) Development of Dream Chaser

In January 2016, NASA awarded one of the Commercial Resupply Services contracts to Sierra Nevada Corporation for six cargo delivery missions to and from the ISS using the Dream Chaser vehicle. This will enable spacecraft reusability and runway landings for US cargo delivery and access to the ISS through 2024. SNC was not selected for the first round of awards but had multiple parties interested in their design and continued with the vehicle's development.

"SNC is honored to be selected by NASA for this critical US program," said Eren Ozmen, chairwoman of Sierra Nevada Corporation. "In such a major competition, we are truly humbled by the show of confidence in SNC and look forward to successfully demonstrating the extensive capabilities of the Dream Chaser spacecraft to the world. SNC's receipt of this award is an American dream come true for all of us. We thank NASA, the Administration and Congress for recognizing the importance of this vital program by supporting the CRS2 contract."[7]

The reusable Dream Chaser vehicle has been in development for over 10 years and offered a space plane that is in some ways similar to the space shuttle—reusable and able to land on a runway.

"The Dream Chaser Cargo System offers NASA a safe, reliable and affordable solution for ISS cargo delivery, return and disposal, ensuring the effective utilization and sustainability of the ISS for years to come," said Mark N. Sirangelo of SNC's Space Systems. "Within a few short years, the world will once again see a United States winged vehicle launch and return from space to a runway landing. We wanted to thank our more than 30 industry, university, international and NASA center partners for helping us make history and open up the next generation of spaceflight (Fig. 2.2)." [8]

[6] Davenport, Christian. Explostions cited in space launch fight. The Washington Post. 22 Sept 2016.

[7] Sierra Nevada Corporation Release: NASA selects Sierra Nevada Corporation's Dream Chaser spacecraft for commercial resupply services 2 contract. 14 Jan 2016. [Internet] [cited 2016 Jan 21]. Available from: http://www.sncorp.com/AboutUs/NewsDetails/2754

[8] Sierra Nevada Corporation Release: NASA selects Sierra Nevada Corporation's Dream Chaser spacecraft for commercial resupply services 2 contract. 14 Jan 2016. [Internet] [cited 2016 Jan 21]. Available from: http://www.sncorp.com/AboutUs/NewsDetails/2754

Fig. 2.2 Sierra Nevada Corporation Dream Chaser pre-drop testing (Image courtesy of NASA)

The Dream Chaser, a snub-nosed version of the space shuttle, was originally developed by a company named SpaceDev (a company with expertise in electronics, avionics and communications systems). Its design was based on an early NASA shuttle design that was itself based on a Soviet spacecraft (HL-20) that never flew. After agreement with NASA to reuse the design, the company struggled to get funding, and after a series of events, SNC purchased the company in 2008. In 2014, SNC was not awarded a contract in the first round of Commercial Resupply Services. However, after further development and focus for flexibility of application, SNC was rewarded with the second round funding. In addition, recent failures by SpaceX and Orbital ATK rockets opened the door for another developer (Fig. 2.3).

The Dream Chaser cargo system includes several features that make it adaptable for multiple types of missions and delivery systems. It is reusable, which lowers costs and allows a quick turnaround for the next mission. A low-g re-entry and pressurized cargo capability allows sensitive payloads to return to Earth safely and quickly for studying microgravity experimental subjects. Other features include a folding wing that allows the spacecraft to fit inside a variety of rockets. Once ready, the Dream Chaser will carry up to seven people to an orbiting complex. It will be launched vertically on an Atlas 5 rocket and, upon return, land on a runway. In support of the Dream

Fig. 2.3 Sierra Nevada Dream Chaser spacecraft and cargo module attached to the ISS. (Image courtesy of NASA)

Chaser development, the European Space Agency is beginning to work on a docking system for the vehicle to attach to the ISS. NASA looks forward to the upcoming tests and certification of the vehicle.

Commercial Crew Program (CCP) Development

As the space shuttle program's end was approaching, the need arose to transport supplies and human payloads to low Earth orbit to the ISS through alternate means. Up until 2010, NASA and the US government did not have any established collaboration with the commercial side of spacecraft development. In mid-2010, the United States released a new National Space Policy that states that "the US Government will use commercial space products and services in fulfilling government needs…[and will] seek partnerships with the private sector to enable commercial spaceflight capabilities for the transportation of crew and cargo to and from the International Space Station."[9]

[9] Federal Aviation Administration. 2011 U.S. Commercial Space Transportation Developments and Concepts: Vehicles, Technologies, and Spaceports.

After the space shuttle program ended in mid-2011, there was a significant business opportunity to meet the demand of low cost transport of cargo and humans to the ISS. NASA has always hired outside contractors to build vehicles or systems to NASA specifications. NASA was now initiating some new ways of performing space business, that is, establish partnerships with corporations that would focus on their own specialties, such as crew vehicles or propulsion, in order to fill the gaps that exist. The result would be greater profits for NASA and increased efficiency.

As previously stated, no other means existed for cargo and humans to be transported to low Earth orbit or to the ISS other than the rockets and spacecraft belonging to other countries, in particular, Russian transport. This was a very controversial situation to American astronauts, scientists, and the US government. To address this problem as quickly as possible, a plan was put in place to develop a US commercial crew transportation capability by 2017. This plan would use commercial partnerships to develop space technology and services while maintaining high NASA standards for safety and reliability. The new program was called the Commercial Crew Program (CCP).[10] Since the implementation of this plan, there has been a boon in the development of space applications of all types—rocket propulsion systems, crew transport vehicles, reusable launch vehicles, life support systems, and satellite delivery mechanisms just to name a few.

NASA awarded more than $8.2 billion under various agreements and contracts[11] including the Space Act Agreements (SAAs) and contracts for different aspects of commercial crew development and transportation. NASA could now focus more on deep space missions while private US companies would develop and operate flights between Earth and the ISS. NASA would have more resources to focus on its future visions of space exploration, leaving the business service of space transport to private enterprise, which would save money while generating American resources and jobs.[12]

NASA's ambitious commercial space program has enabled a successful partnership with two American companies to resupply the International Space Station. A little more than 2 years after the end of the space shuttle program, SpaceX and Orbital ATK began successfully resupplying the space station with cargo launched from the United States. The companies developed the rockets and spacecraft through public-private partnerships under the agency's

---

[10] Progressive Media—Company News. NASA's CCiCap programme completes first milestone. 27 Aug 2012.

[11] Nasa.gov. Commercial crew program—the essentials. [Internet] NASA.gov; [cited 2016 Mar 23]. Available from:
https://www.nasa.gov/content/commercial-crew-program-the-essentials/#.VvMWQuIrK70

[12] Anderson, Chris. Nov 2013. Rethinking public-private space travel. Space Policy. 29:4:266–271.

Commercial Orbital Transportation Services (COTS) program, an initiative that aimed to achieve safe, reliable and cost-effective commercial transportation to and from the space station and low-Earth orbit. NASA then awarded Orbital ATK and SpaceX commercial resupply services contracts to each to deliver at least 20 metric tons of cargo to the orbiting laboratory.

Early in 2010, NASA awarded $50 million in stimulus money to support five companies in developing technologies for upcoming space missions. The companies would provide matching funds from other sources.[13] The companies and their development and accomplishments since that time are:

- *Sierra Nevada Corp*: $20 million for the development of the Dream Chaser lifting body space plane, designed to launch vertically and land on a runway, inspired by NASA lifting body designs. Their vehicle lost out in the competition for NASA's commercial cargo transport to Space Exploration Technologies Corp. and Orbital Sciences Corp. Despite this setback, the Dream Chaser program publicly offered Dream Chaser vehicles for purchase to the European Space Agency and the Japan Aerospace Exploration Agency, who had recently expressed interest in them. In addition, a remodeled craft entitled the Dream Chaser cargo system will be submitted for the next NASA Commercial Resupply Services contract.[14]
- *Boeing Co*: $18 million to develop a spacecraft that is compatible with a variety of launch vehicles. Boeing had extensive experience with the development of systems designed for human use, particularly with the ISS. In April 2011, Boeing was selected for the second round of the crew development program and will continue refining its Crew Space Transportation (CST)—100 spacecraft. Crew rotation mission flights are scheduled for late 2017, provided the craft is ready and the funding remains stable. The CST-100 is capable of carrying up to seven people or the equivalent space for a combination of humans and cargo.[15]
- *United Launch Alliance* (*ULA*): $6.7 million to develop a launch system for the Boeing CST-100 to the ISS. ULA represents the joint venture of Lockheed Martin and Boeing. Initial test flights are planned for the ULA Atlas launch vehicle in 2015.
- *Blue Origin*: $3.7 million to develop a new American-built rocket engine, the BE-4, to power the ULA vehicles. The rocket will meet both commercial and military requirements.

[13] Clark, Stephen. NASA selects winners of first commercial crew contest. Spaceflight Now. 02 Feb 2010.
[14] Keeney L. 2014 was the year of Colorado space exploration. Denver Post. 28 Dec 2014.
[15] PR Newswire. Boeing selected for 2nd round of NASA commercial crew development. 18 Apr 2011.

- *Paragon Space Development Corp*: $1.4 million to provide environmental controls for commercial human spaceflight systems. [16]

SpaceX was missing from the awards most likely because funding was earmarked for developing technologies, whereas SpaceX was already much further along in the development of the *Falcon 9* and Dragon spacecraft used to ferry cargo to the ISS.

In April 2011 NASA distributed its second group of awards—$269 million to four companies to develop spacecraft to take American astronauts to orbit more quickly and less expensively than humans in the past. Several of the same corporations received further funding to spur on their final development:

- *Space Exploration Technologies Corporation* (*SpaceX*): $75 million towards adapting the Falcon 9 rocket and the Dragon capsule for human flight. The vehicles have had a number of successful flight tests since that time and a few setbacks. A detailed account of SpaceX was given earlier.
- *Sierra Nevada Corp*: $80 million to further develop the Dream Chaser, which was not chosen to be in the final group but had several other commercial options—see their story earlier.
- *Boeing Co*: $92.3 million to continue the capsule design.
- *Blue Origin*: $22 million to develop its capsule. See their story above.

All of the spacecraft being developed, with the exception of the SpaceX vehicle, need a launch system. The United Launch Alliance's Atlas V rocket is still looking as the top choice for this application.[17] Bigelow Airspace recently announced a deal to use the Atlas V in launching an expandable habitat into orbit in 2020, and more such ventures are expected.

In September 2014, NASA chose Boeing and SpaceX to receive the final awards to finish the development, testing and certification for their spacecraft and commercial crew transportation services.[18] By mid-June 2015, the Senate subcommittee considering the NASA budget approved $900 million (the House version appropriates $1 billion) for the commercial crew program. NASA says that $1.2 billion is needed to keep the program fully funded. This reduction would result in launch delays of at least another 2 years of the first flights of the Boeing and SpaceX capsules, pushing them past 2017, requiring

[16] Clark, Stephen. NASA selects winners of first commercial crew contest. Spaceflight Now. 02 Feb 2010.

[17] Chang, Kenneth. NASA awards $269 million for private projects: [national desk]. The New York Times. 19 Apr 2011: A.13.

[18] Science committee democrats congratulate Boeing and SpaceX on NASA's commercial crew development awards. Targeted News Service. 16 Sep 2014.

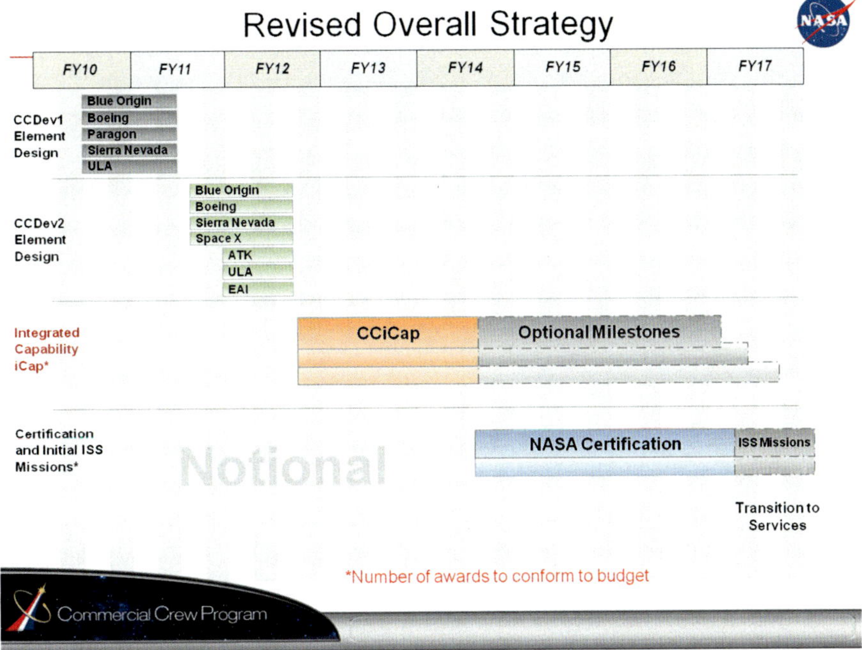

Fig. 2.4 Commercial crew schedule (Image courtesy of NASA)

further reliance on Russian spacecraft and rockets. In addition, NASA reports spending $500 million per year having our astronauts ferried per agreements with the Russian space agency.[19] The message was heard and the commercial crew program was given over $1.24 billion in the 2016 budget.[20]

There are currently daily advancements and funding news updates in the development and testing of the commercial crew vehicles and launch systems with an ambitious schedule and critical 2016 milestones leading originally to a 2017 completion date (Fig. 2.4). In May 2016, Boeing announced a delay of the first crewed flight until early 2018 due to technical problems as well as adjustments affecting the program. Once operational, NASA stated, the vehicle would be capable of making two crew launches a year from the United States to the International Space Station.[21]

[19] Clark, Stephen. Commercial crew spaceships face likely delays. Spaceflight Now. 10 June 2015.

[20] Grush, Loren. Congress wants to give NASA $19.3 billion next year, even more than Obama asked for. The Verge. 16 Dec 2015. [Internet] [cited 2016 Jan 21]. Available from: http://www.theverge.com/2015/12/16/10289030/nasa-budget-increase-2016-congress-funding

[21] NASA.gov. NASA ordersSpaceX crew mission to international space station. [Internet] NASA.gov Release 15-240; Dec. 18, 2015 [Internet] NASA.gov Release 15-240; Dec. 18, 2015 [cited 2016 Mar

"Once certified by NASA, the Boeing CST-100 Starliner and SpaceX Crew Dragon each will be capable of two crew launches to the station per year," said Kathy Lueders, manager of NASA's commercial crew program. "Placing orders for those missions now really sets us up for a sustainable future aboard the International Space Station."[22]

Commercial crew missions to the ISS will restore America's manned spaceflight capabilities and increase scientific research of both Earth and outer space. A standard commercial crew mission to the station will carry up to four crew members and about 220 pounds of pressurized cargo. The spacecraft will remain at the station for up to 210 days, also available as an emergency lifeboat during that time.[23]

In early 2016, NASA awarded its next round of Commercial Resupply Services contracts to continue its partnership with commercial companies for cargo resupply of the ISS. Boeing's unscrewed version of the CST-100 Starliner was eliminated from the group. SpaceX and Orbital ATK won contracts to continue developing their Dragon and Cygnus vehicles. In addition, Sierra Nevada Corporation's Dream Chaserwas added to the list of commercial cargo resupply vehicles delivering to the ISS.

"This second CRS contract award reinforces Orbital ATK's role as a trusted partner to NASA with a proven cargo delivery and disposal service that continues to support the important work being performed aboard the ISS," stated Frank Culbertson, president of Orbital ATK's Space Systems Group.[24]

In July 2015, NASA named four of their most experienced astronauts to train to fly orbital flight tests in SpaceX's Dragon spacecraft and Boeing's spacecraft (CST-100 Starliner). These astronauts will be the first to train for commercial spaceflights that will return launch capabilities to the United States. The first group of astronauts selected, all veterans, are already participating in an intensive training program to prepare for the 2017 launch of the CST-100 spacecraft.[25] "These distinguished veteran astronauts are blazing a new trail—

23].Availablefrom:http://www.nasa.gov/press-release/nasa-orders-spacex-crew-mission-to-international-space-station

[22] NASA.gov. NASA orders second Boeing crew mission to international space station. [Internet] NASA.gov Release 15-240; Dec. 18, 2015 [cited 2016 Mar 21]. Available from: http://www.nasa.gov/press-release/nasa-orders-second-boeing-crew-mission-to-international-space-station

[23] NASA.gov. NASA orders second Boeing crew mission to international space station. [Internet] NASA.gov Release 15-240; Dec. 18, 2015 [cited 2016 Mar 21]. Available from: http://www.nasa.gov/press-release/nasa-orders-second-boeing-crew-mission-to-international-space-station

[24] Gebhardt, Chris and Bergin, Chris. NASA awards CRS2 contracts to SpaceX, Orbital ATK, and Sierra Nevada. Nasaspaceflight.com [Internet] Jan. 14, 2016 [cited 2016 Mar 21]. Available from: https://www.nasaspaceflight.com/2016/01/nasa-awards-crs2-spacex-orbital-atk-sierra-nevada/

[25] NASA.gov. NASA selects astronauts for first U.S. commercial spaceflights. [Internet] NASA.gov Release 15-148; Jul. 9, 2015 [cited 2016 Mar 21]. Available from:

http://www.nasa.gov/press-release/nasa-selects-astronauts-for-first-us-commercial-spaceflights-0

a trail that will one day land them in the history books and Americans on the surface of Mars," said NASA administrator Charles Bolden.[26]

Space Contests and Prizes

XPRIZE is a non-profit organization that formulates and conducts public competitions that require technological development representing "radical breakthroughs for the benefit of humanity." The prize concept was used previously in aviation, with Charles Lindbergh winning the second Orteig Prize in 1927, a $25,000 reward offered by a New York hotel owner to the first aviator to fly non-stop from New York to Paris or the reverse. The XPRIZE organization's high profile trustees include Elon Musk, James Cameron, Arianna Huffington, and others.[27]

This concept of an incentivized race to achieve a particular set of technological goals inspired the first XPRIZE. In 1996, entrepreneur Peter Diamandis offered a $10 million prize (funded by the Ansari family) to the first privately funded team that could build and fly a three passenger reusable vehicle that could travel 100 km into space twice within 2 weeks. The contest was later titled the Ansari XPRIZE. The prize was awarded in 2004 to Mojave Aerospace Ventures (a company backed by Sir Richard Branson) with their spacecraft SpaceShipOne, designed by Burt Rutan.[28]

Many consider the successful flight of SpaceShipOne as the beginning of a new era for private enterprise commercial options for outer space. Things progressed for Mojave Aerospace Ventures with some success and also setbacks. SpaceShipTwo crashed during a test flight Oct. 31, 2014, killing the copilot and injuring the pilot. It was determined that the accident was due to pilot error.[29] The company and Sir Richard Branson were devastated and considered ending the venture. However, the company pulled together, and SpaceShipTwo is being redesigned and rebuilt, with the eventual goal of taking passengers on short trips to space.

The next great XPRIZE contest, the Google Lunar X Prize, was a Moon challenge sponsored by Google Inc. in 2007, offering up to $30 million in an

---

[26] NASA picks four astronauts to fly first commercial space mission. Shanghai Daily. 2015 Jul 10.

[27] Shivapriya, N. All for a good cause XPRIZE can help solve big problems: Peter Diamandis. Economic Times (E-paper edition). 12 Dec 2014.

[28] Hudgins, Edward. Signals from SpaceShipOne. Washington Times. 06 Oct 2004. A14.

[29] Harwell, Drew & Brenner, Glenn. Spacecraft crash shows test-pilot dangers. The Washington Post. 06 Nov 2014. A20.

array of prizes. The first private company to launch from Earth, land a robotic rover on the Moon, travel at least 500 m on the surface, and transmit high definition data and video footage to Earth by end of 2017 will take home $20 million. Bonus prize money would be awarded for various achievements such as length of travel, operating at night, detecting water, and accuracy of landing at a particular location. Five teams were already awarded money for some of these prizes—Astrobotic (USA), Hakuto (Japan), Indus (India), Moon Express (USA) and Part-Time Scientists (Germany). The international participation is remarkable—all continents are represented except for Africa. At least one team must announce a launch date before the end of 2016; otherwise, the competition will expire. There were still 26 teams involved in the primary race by the end of 2011. Due to delays associated with technology and competitor financing, the deadline was extended to the end of 2016 from an original earlier end date.[30]

Another $50 million lunar competition, the O-Prize or American Space Prize, was offered by hotel magnate Robert Bigelow of Bigelow Aerospace for the first American team to launch a manned spacecraft into orbit by 2010. The date came and went without a winner. Bigelow Aerospace, however, has been a pioneer in considering space tourism and travel. The company is featured below.

Inflatable Habitats

Bigelow Aerospace is owned by Robert Bigelow, a real estate investor and space entrepreneur. He earned his fortune in the hotel business (Budget Suites) and is interested in extending tourism into outer space, focusing on inflatable space station modules as a start. Bigelow has visions for both space stations and a Moon base. He currently has two inflatable prototype modules in orbit—Gemini 1 (launched July 2006) and Gemini 2 (launched June 2007). The company will observe how these modules perform long term (Figs. 2.5, 2.6 and 2.7).

Bigelow Aerospace has been developing expandable modules (Bigelow Expandable Activity Module) designed to attach to the International Space Station (ISS). The first module was delivered by the SpaceX's Dragon supply ship launched aboard a Falcon 9 rocket in March 2016. The module was then attached to the ISS and successfully inflated after some minor technical difficulties.[31] Figure 2.5 illustrates the module expansion after attachment to the

[30] Yeld J. 26 teams in race to reach moon first and strike it rich. The Argus. 2011 Oct 05; Sect. 4.

[31] Clark, Stephen. Dragon arrives at space station with inflatable habitat in tow. 10 Apr 2016. [Internet] [cited 2016 Jun 26]. Available from: https://spaceflightnow.com/2016/04/10/dragon-arrives-at-space-station-with-inflatable-habitat-in-tow/

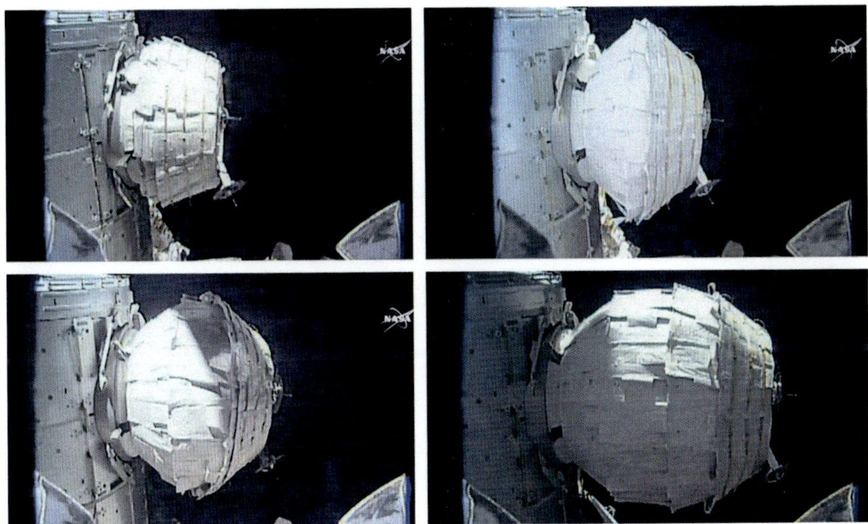

Fig. 2.5 The Bigelow Module Expansion Steps after Attachment to the ISS. (Image courtesy of NASA TV)

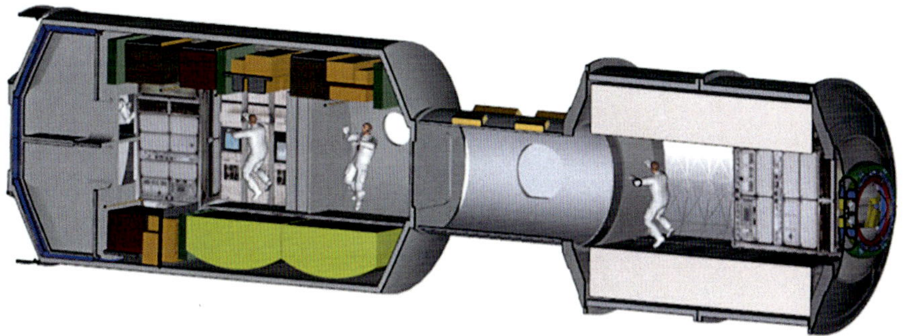

Fig. 2.6 Artist's concept of deep space living module (Image courtesy of NASA)

ISS. NASA awarded a $17.8 million contract to Bigelow to design and manufacture the habitat. These habitats could eventually be the prototypes of the first human habitats on the Moon, and being lightweight and compact, both Bigelow and NASA foresee using them for Mars expeditions. They fold up into a small space and inflate like an air mattress. Segments can be connected to form bigger modules, creating the possibility of future space hotels. The modules will be monitored for a period of 2 years to test their durability while measuring life threatening space radiation and micrometeorite strikes.[32]

[32] Maynard, James. Bigelow Aerospace to usher in future of ISS with expandable space station. Tech Times. 13 March 2015.

Fig. 2.7 The Bigelow Expandable Activity Module (BEAM) Attached to the ISS. (Image courtesy of NASA)

3

Mars

Keywords Asteroid Redirect Mission (ARM) • Buzz Aldrin • Charles Bolden • Constellation Moon • Elon Musk • George Abbey • Hohmann Transfer • Japan Aerospace Exploration Agency (JAXA) • Laurel Kaye • Life Support System (LSS) • Low-Density Supersonic Decelerator (LDSD) • Lunar CATALYST Initiative • Mariner spacecraft • Mars • Mars One • Mars Science Laboratory (Curiosity) • Martian atmosphere • National Research Council (NRC) • Opportunity spacecraft • Orion spacecraft • President George H.W. Bush • President George W. Bush • Rare Earth elements (scandium, lanthanum, cerium) • Rover spacecraft • Saturn V • Skycrane maneuver • Smart-1 • Sonia Van Meter • Space Exploration Initiative (SEI) • Space Launch System (SLS) • SPACEX FH LIFT ROCKET • Spirit spacecraft • Supersonic Inflatable Aerodynamic Decelerator (SIAD) • Viking probe • XPrize • Zubrin

> "By refocusing our space program on Mars for America's future, we can restore the sense of wonder and adventure in space exploration that we knew in the summer of 1969. We won the Moon race; now it's time for us to live and work on Mars, first on its moons and then on its surface."[1]
>
> –Buzz Aldrin (2009)

Little green men, the Red Planet, stark but beautiful landscapes... We all have our own images in our mind of what Mars is like and opinions about whether Mars should be the next big step in manned space exploration. Those of us who are lovers and advocates of space travel support confronting the

[1] Aldrin, Buzz. Commentary: let's aim for Mars. CNN.com. 23 June 2009. [Internet] [cited 2016 Feb 12]. Available from: http://www.cnn.com/2009/TECH/space/06/23/aldrin.mars/

L. Dawson, *The Politics and Perils of Space Exploration*, Springer Praxis Books, DOI 10.1007/978-3-319-38813-7_3

challenges that such a trip would present. Among the supporters are scientists and astronauts who feel that the United States is losing its dominance in space travel, both manned and unmanned. Buzz Aldrin, the second man to walk on the Moon, believes that America needs strong leadership to inspire the nation to make strides in space travel. "America must be the world leader in human spaceflight," he said in early 2015. "There is no other area that clearly demonstrates American innovation and enterprise more than human space flight."

Buzz Aldrin outlined a "unified space vision," a plan for American space exploration and the colonization of Mars. His focus is to build enthusiasm among the young and old to travel into deep space. As we approach the 50th anniversary of walking on the Moon (July, 2019), he is hoping that America commits to colonizing Mars and continuing scientific research in deep space.

"Humans need to explore, push beyond current limits just like we did years ago," Aldrin said. "Apollo was the story of people at their best, working together for a common goal. We started with a dream, and we can do these kinds of things again. I know it. I'm living proof that it can be done."

Aldrin wants to establish a permanent residence on Mars by 2040, with preliminary missions beginning before 2020. The plan is outlined in his book *Mission to Mars: My Vision for Space Exploration*. The plan is based on a concept called "The Cycler," a spacecraft that continually travels between Earth and Mars.

"People ask me sometimes 'Why do we need to go to Mars?'" Aldrin said. "'Why do we even need a space program?' Because by adventuring into space, we improve life for everyone here on Earth. The scientific advancement, the innovations that come from this kind of research, new technology that we use in our daily lives..."[2]

Others would argue that we might be better off spending our resources on causes more important than space exploration. The fact is that the NASA budget has only been a flat 0.50 % of the entire US federal budget for the past 5 years.[3] In addition, some of our most important goals surrounding making life better here on Earth have to do with sustainability, development of transportation that does not rely on fossil fuel, and the use of advanced technologies for medical advancement. These issues could be addressed by deep space exploration and the establishment of a colony on Mars. Some of the biggest scientific breakthroughs in the past decades have come from spinoffs of space research. Investment in space brings about benefits for all humankind.

---

[2] NASA.gov News. May 15, 2015. [Internet] [cited 2015 June 02]. Available from: http://www.nasa.gov/news/budget/index.html

[3] Kuta, S. Apollo astronaut buzz aldrin lays out plan for mars colonization in talk at CU-boulder. The Daily Camera. 2015 Mar 03.

Finally, deep space travel and eventually the colonization of Mars will provide options for our survival. If we can spread civilization out throughout the solar system and the universe, we will have a better chance should a disaster strike our planet.

A Dangerous High Tech Landing on Mars (August 6, 2012)

The pod containing the rover Curiosity approaches the Martian atmosphere (about 78 miles above the planet's surface to start) with a blazing velocity of 13,000 mph (21,000 km/hour). The atmosphere increases the friction against the spacecraft as it descends toward the surface, resulting in scorching high heat (3800 °F or 2100 °C) on the outside of the pod, which is protected by a heat shield. The journey was a long one—about 352 million miles over 8 months. The NASA's Mars Science Laboratory (MSL), also referred to as the rover Curiosity, is tucked inside of a larger spacecraft that protects it from the outside elements. In a short period of time, the craft needs to slow from 13,200 to 1000 mph (5.9 to 0.45 km/second). A crash would end a $2.5 billion mission and maybe the future of Mars human exploration.

At 7 miles (11 km) a giant parachute—the biggest chute ever used in an outer space landing at 51 feet (15.5 m)—deploys to slow the spacecraft to a gradual descent. About 5 miles (8 km) above the surface, the heat shield comes off and drops away. Shortly after, the radar determines when to start powered descent. As with landing on the Moon, the powered descent needs to be started exactly at the right time, otherwise, the craft will run out of fuel (start too early) or will still be descending too fast without enough time to stop (start too late). About a mile (1.6 km) from the surface, the protective outer shell drops the MSL. The eight retrorockets fire and allow it to free fall and fly away from the descent stage. The craft has been slowed to about 170 mph (274 km/hour).

The spacecraft performs the final tricky maneuver known as the sky crane maneuver, lowering the MSL to the surface using cables which lower about 25 feet (7.5 m). Curiosity is lowered slowly and then separated from the descent vehicle. At final separation, descent has been slowed to less than 1.7 mph (2.7 km/hour). Curiosity's wheels and suspension system double as landing gear and are deployed close to the surface. The cables are spooled as the rover descends and softly touches down. The descent stage knows the rover has landed when the weight on the cables is reduced. The bridle and cord is cut and a connection to the computer brain is severed, and the sky

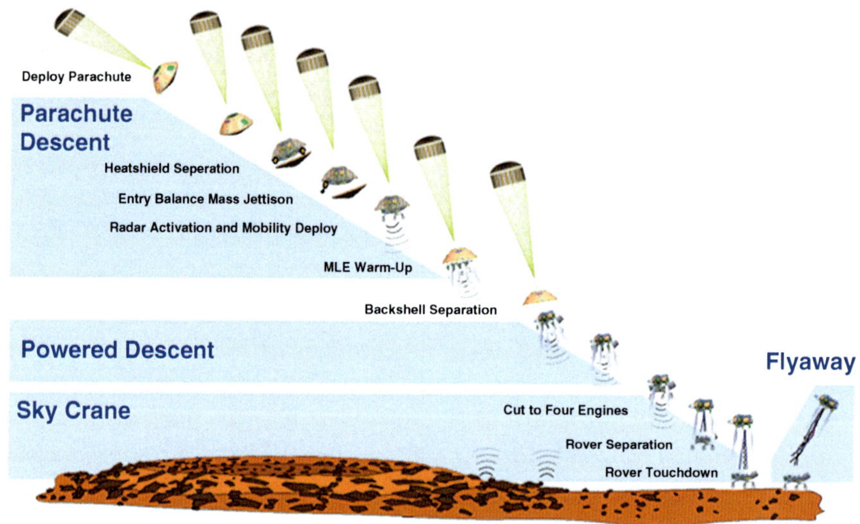

Fig. 3.1 Mars Science Laboratory (Curiosity) entry and descent to Mars sequence (Image courtesy of NASA)

crane flies away. The descent stage flies out of the way, almost 500 feet (about 150 m) from the rover. The nuclear-powered rover Curiosity moves on to fulfill its scientific research and preparation for further missions (Fig. 3.1).[4]

NASA described this delicate and untried landing method as "seven minutes of terror." This complex landing maneuver is one example of the demonstrated technology and ingenuity required to send heavy equipment and eventually humans to Mars. The success of the MSL landing is another testament to NASA's creativity and expertise in space exploration. The entry method included technology from past NASA Mars missions along with new technologies. Instead of the airbag landing of past Mars missions, the Mars Science Laboratory used a guided entry and a sky crane maneuver to land the massive Curiosity rover (over 2000 pounds, or 900 kg). The new aspects of the guided entry made the landing more precise, making specific target areas more possible for future missions. Curiosity landed about 1.5 miles from its destination, the Gale Crater.[5]

[4] NASA.gov. [Internet] [cited 2015 June 02]. Available from: http://mars.nasa.gov/msl/mission/timeline/edl/

[5] Space.com Staff. 5 June 2014. Innovation the NASA way (US 2014): book excerpt. Space.com. [Internet] [cited 2015 June 05]. Available from: http://www.space.com/26135-innovation-nasa-book-excerpt.html

Over the past couple of years, Curiosity has been determined to be a success, still roaming the surface remotely controlled and transmitting data back to Earth. It has survived radiation, dust storms and other hazards on the Martian surface. As for discoveries, it has found elements supportive of life, evidence of water but no organic molecules so far. In early 2015, the robotic rover detected nitrogen on the Martian surface in the form of nitric oxide, a product of nitrates that can be used by living organisms. In addition, Curiosity has found that water can exist as a liquid near the planet's surface.[6] This finding leads scientists to support the theory that dark streaks seen on crater walls could have been formed by flowing water. Although there is nothing definitive for now, the discovery is another piece of the puzzle to determine the presence of past or present life on Mars.[7]

The Case for Mars

Mars, the fourth planet from the Sun, holds a certain fascination for those of us who grew up with science fiction movies, Orson Welles and the "War of the Worlds," and the vision of little green men. We can now actually gaze at the Red Planet and wonder what it would be like to travel or live there—something that could be accomplished in our lifetime. These same science fiction books and movies have fueled the quest for life on Mars and encouraged investigation into the existence of life supporting elements that could have connections to the beginnings of life on Earth. New discoveries on the surface of Mars are providing evidence that water once was plentiful on the Martian surface, one of the primary requirements for past life on Mars. Although remarkable, it is important to note that the existence of water is just one of the requirements for life as we know it. It is thought that Mars was hospitable enough to at least support a primitive type of bacteria in the past, a far cry from little green men.

Robotic rovers have explored Mars over the past few decades, transmitting numerous scientific information and visual images about the planet. However, it will take more complex human missions to explore Mars' numerous surface features and to drill into subsurface groundwater. Mars appears to be the next logical step in human space travel. It is more hospitable and accessible than other planets, asteroids or moons. The Martian surface also has enough space to accommodate a human colony and possesses a variety of materials that can be mined and processed to build structures and equipment.

[6] Mars has nitrogen, key to life: NASA. The Nation [Bangkok]. 2015 Mar 25.

[7] Martin-Torres, Javier et al. May 2015. Transient liquid water and water activity at Gale crater on Mars. Nature Geoscience. 8:357–361. Published online: 13 April 2015.

Table 3.1 Comparison of Earth and Mars (http://mars.nasa.gov/allaboutmars/facts/#infographic)

	Earth	Mars
Average distance from Sun	93 million miles	142 million miles
Average speed in orbiting Sun	18.5 miles per second	14.5 miles per second
Diameter	7926 miles	4220 miles
Tilt of axis	23.5°	25°
Length of year	365.25 days	687 Earth Days
Length of day	23 hours 56 minutes	24 hours 37 minutes
Gravity	2.66 times that of Mars	0.375 that of Earth
Temperature	Average 57 °F	Average −81 °F
Atmosphere	Nitrogen, oxygen, argon, others	Mostly carbon dioxide, some water vapor
Number of moons	1	2

Scientists have eliminated Venus, the second planet from the Sun, for colonization. At first look, it has some potential for recreating our own environment. It is close to the same size as Earth, giving it a similar gravitational field. That, however, is where the similarity ends. Surface pressures are so high that they would crush us to death in an instant. The extremely high surface temperatures would be prohibitive for humans to exist. Add to that sulfuric acid clouds and rain, and Venus as a human destination was scrapped as a near future option by all space agencies.

It is thought that we can reach Mars with a minimal extension of existing technology. Once there, if we want to live on the planet for any length of time, humans will have to build a sustainable environment using solar and other types of energy to maintain life support. Recently, astronauts grew and ate lettuce on the International Space Station, a step to better understanding of how to produce food crops outside of Earth. In addition, an astronaut (Scott Kelly) and a cosmonaut (Mikhail Kornienko) spent 1 year on the ISS providing valuable data on human physiology and psychology for long duration spaceflights.[8] Progress is definitely being made also in technology and transportation systems to make the Mars mission a reality.

This chapter focuses on the challenges to colonizing Mars and two current approaches to Mars exploration—the NASA mission to Mars and the Mars One project.

To provide some background, Table 3.1 provides some of the most common comparisons between Earth and Mars (Fig. 3.2).

[8] Space.com Staff. 5 Sep 2015. A manned mission to Mars is closer to reality than ever. Space.com. [Internet] [cited 2015 June 05]. Available from: http://www.space.com/30580-nasa-manned-mars-mission-reality.html

Fig. 3.2 Earth and Mars side by side (Image courtesy of NASA)

The two terrestrial planets, Earth and Mars, have many similarities such as comparable rotational axis tilts (resulting in seasonal variability), similar length of a day, and the existence of polar caps. Some of the differences, such as a low gravitational force on the surface of Mars, about one third of that on Earth, will provide a challenge for human colonization. The effects of a low gravity environment on humans include loss of bone density and muscle mass over time. It may be necessary to introduce an artificial gravity environment, although there are no plans for this kind of facility at this time and no definite solutions to this problem for long-term residency. Another challenge for humans on Mars is the thin protective atmosphere that surrounds Mars with a composition that is almost entirely carbon dioxide. The air is not only unbreathable but provides little protection against radiation from the Sun. (Dangers of humans in outer space are further discussed in Chapter 5.) Space suits would be required for outside work or travel on the surface of the planet. Exposure to radiation would be limited to short periods of time; living and working modules would have to be shielded.

Another characteristic of the planet requiring protective clothing or a suitable habitat is the average surface temperature, which is –81 °F on Mars versus 57 °F for Earth. Seasonal variation can result in even harsher temperatures. Dust storms make landers, rovers and modules susceptible to tipping over in strong winds, and blowing dust particles can infiltrate electronics and other hardware, affecting performance.

Yet, Mars possesses all of the required raw materials to support life and has enough land to create a large human colony. Understanding Mars can tell us many things about our own planet and its origins as well as teach us about the formation and structure of other worlds in our Solar System.

As an exploration target, Mars remains high on the list of importance for scientists of many disciplines. A couple of the most vocal supporters of Mars colonization are discussed next.

The Supporters of Mars Colonization

Robert Zubrin

Robert Zubrin (1952–) is the founder and president of the Mars Society (http://www.marssociety.org), an organization whose purpose is to further the exploration and settlement of the Red Planet using both public and private means. He is also president of Pioneer Astronautics, an aerospace R&D company in Colorado. He has degrees in aerospace engineering and a Ph.D. in nuclear engineering from the University of Washington. He is currently one of the advisors for the Mars One mission, an international one-way human colonization effort.

Dr. Zubrin is also an inventor and scholar in space propulsion and strategies for space travel. While working for Lockheed Martin, he developed a "Mars Direct" mission plan that uses Martian resources to reduce a manned Mars exploration program to about an eighth of what was previously estimated by NASA. The Mars Direct plan is a minimalist approach to exploring Mars using existing launch technology to travel there and Martian resources to generate rocket fuel (using the abundance of carbon dioxide processed into methane for propellant), extract water and construct living modules. The mission's approach is to establish a colony by rotating crews that build on each other's success to create a step-by-step complete colony.

Zubrin believes that the US space program needs a challenge. "A bold humans-to-Mars program would also be a challenge to every kid in the country: Learn your science, and you can help explore a new world. Imagine the potential that an ambitious Mars program could have in inspiring legions of future engineers, inventors, medical researchers, doctors and other scientists."

Zubrin's plan is detailed in his popular and timeless book: *The Case for Mars: The Plan to Settle the Red Planet and Why We Must* (Free Press, Second Edition, 2011). Years after publication, the book still remains informative and inspirational as a vision for space exploration and provides a focus for future endeavors.

"Mars" is the new world. Its settlement presents the challenge that will determine whether we remain confined to Earth, or can become a multi-planet spacefaring species, with a future made unbounded by our courage and creativity. Mars One has accepted that challenge. It is a daunting one, and the odds may well be against them. But if no one tries, no one will succeed. I'm proud to do what I can to help."[9]

Elon Musk

Elon Musk (1971–) is the CEO of SpaceX and the CEO and product architect of Tesla Motors. He is an entrepreneur, inventor, engineer and investor, overseeing rocket and spacecraft development to Earth orbit and beyond. In 2008, SpaceX's *Falcon 9* rocket and Dragon spacecraft won the NASA contract to provide a commercial replacement for the cargo transport function of the space shuttle. In 2012, SpaceX became the first private company to dock with the International Space Station and return cargo to Earth using the Dragon. The company specializing in small satellite launches and cargo runs is now moving into the business of human spaceflight.

Musk detailed his vision and plans for colonizing Mars in early 2016. SpaceX will send an unmanned mission to Mars on its Flying Dragon version 2 rocket starting in 2018 with continuing launches approximately every 26 months. If these flights go well, their first rocket carrying human payloads will launch in 2024. The vehicle has the capability to carry up to seven people for the 18-month trip. It does not have the capability to return to Earth, so the plan is for colonization.[10] He has argued that we must put a million people on Mars if we are to ensure that humanity has a future. He believes that going to Mars is as urgent an issue as eradicating poverty and disease.

It is true that human civilization won't last long in this universe if it stays on a single planet. We know that our Sun will one day grow so large it will destroy Earth. Even though it will take 5–10 billion years, there will be significant changes prior to that event, as soon as 500 million years from now.[11] If humanity is to survive, it will need to expand outside of Earth's system. Setting up a Mars colony is one step in that direction to being independent of Earth.

Musk is an enthusiastic supporter of space exploration. After a perceived American loss of space vision, he began planning a Mars mission of his own.

[9] Zubrin, Robert. The Promise of mars. May/June 1996. Ad Astra. [Internet] [cited 2015 May 19]. Available from: http://www.nss.org/settlement/mars/zubrin-promise.html

[10] Taylor, Harriet. [Internet]. CNBC.com; c2016. Musk: we intend to launch people to Mars in 2024. 02 Jun 2016. CNBC.com. [cited 2016 Jun 28]; Available from: http://www.cnbc.com/2016/06/02/musk-we-intend-to-launch-people-to-mars-in-2024.html

[11] Eicher D. Feb 2013. How long can we last on Earth?. Astronomy. 41:2:7.

Musk funded his own rocket company and now has visions of a self-sustaining civilization on another planet. He sees the trip as one-way, and the first group of colonists would have to pay their own way. Far into the future, he believes, as does Zubrin, that Mars could be terra-formed into Earth-like surroundings. At the start, the habitat would be crude and challenging, and those that survive for decades or longer will be the new settlers, pioneers. In order to form a genetically diverse civilization, Musk believes that it would take a million people, several hundred trips of a giant spaceship (which he calls a Mars Colonial Transporter). This begins to sound like science fiction, but it is comforting in a way to know that a lot of thought has been put into our survival.

Musk says, "If we're going to have any chance of sending stuff to other star systems, we need to be laser-focused on becoming a multi-planet civilization. That's the next step."[12]

Unmanned Mars Exploration

There have been over three dozen unmanned missions launched from six countries to help us understand more about Mars. Some were only brief flybys (early 1960s), gathering small amounts of data and visually scanning a number of different geographic regions. Other missions lasted for years and involved orbiters and landers that gathered a significant amount of information needed to answer more detailed scientific questions. The organizations and countries that have invested in Mars missions are NASA (the United States), the Soviet Union, the European Space Agency, and the Indian Space Research Organization. The United States and Soviet Union were the only two countries initially investing in a number of early missions, only a few years after Sputnik was launched in 1957. Many of the early flights were failures, but hundreds of visual images were sent back to Earth for analysis. The Soviet Union stopped their Mars missions in the early 1980s.[13] Early Mars missions and their discoveries can be seen on the NASA.gov and space.com websites.

One mystery that these early flights solved was the idea that canals sited on the surface of Mars provided possible evidence of intelligent life on the surface. These canals were first seen by the Italian astronomer Schiaparelli in 1877 and later by more sophisticated observatories. Debates continued about

[12] Anderson, Ross. Exodus. Sept 30, 2014. Aeon Magazine. [Internet] [cited 2015 May 19]. Available from: http://aeon.co/magazine/technology/the-elon-musk-interview-on-mars/

[13] Tate, Karl. [Internet]. Space.com; c2014. Mars curiosity: facts and information; Feb 27, 2013 [cited 2015 June 04]. Available from: http://www.space.com/16575-mars-exploration-robot-red-planet-missions-infographic.html

whether the canals were real or were optical illusions until the NASA Mariner spacecraft visited Mars on the 60s and 70s and proved that the canals did not exist. The Viking missions in the 1970s had involved both orbiters and then landers that collected soil samples. When analyzed, the result showed no evidence of life.[14]

In the early 2000s, the Spirit and Opportunity rovers:

- measured the atmosphere of Mars and detected argon,
- explained how dust and sand are moved by wind in Mars' very thin atmosphere,
- demonstrated recent water or frost altering rock surfaces, identified an ancient hydrothermal system,
- and found evidence that Mars may have been capable of supporting life for millions of years.[15]

Recent significant discoveries have been made by the rover Curiosity, launched in 2011 and successfully landing in 2012 after completing a complicated set of maneuvers. Curiosity is the most sophisticated unmanned vehicle yet sent to Mars, with the ambitious mission of finding out if Mars can support life.

Curiosity drilled samples and analyzed the composition to include the fundamental elements that could support life: sulfur, nitrogen, hydrogen, oxygen, phosphorus and carbon. In addition, Curiosity identified organic materials considered to be the building blocks of life. These results point to a suitable environment to support life.[16]

"A fundamental question for this mission is whether Mars could have supported a habitable environment," stated Michael Meyer, lead scientist for NASA's Mars Exploration Program. "From what we know now, the answer is yes."[17]

Radiation levels remain a concern for long term flight and colonization. Curiosity's instruments detected levels that were acceptable for the length of a Mars mission. Long-term exposure is still an issue that would need to be dealt with in the design and structure of living modules as well as protective space suits.[18]

---

[14] Quest to find life beyond Earth never-ending dream. The Daily Telegraph (Australia). 2015 Jul 23.

[15] Petit, Charles W. 2004 Sep 13. Roving about the red planet. U.S. News & World Report. 137:52–54.

[16] NASA rover results include first age measurement on mars and help for human exploration. U.S. Newswire. 2013 Dec 09.

[17] Sample I. Was there life on mars? maybe, say astronomers: Strongest evidence yet of past life on red planet curiosity rover finds signs of ancient river network. The Guardian. 2013 Mar 13; Sect. 5.

[18] Howell, Elizabeth. [Internet]. Space.com; c2014. Mars curiosity: facts and information; March 26, 2015 [cited 2015 April 22]. Available from: http://www.space.com/17963-mars-curiosity.html

Life From and on Mars

Discussions surrounding the mystery of life originating on Earth from Mars were reborn in 1996 when NASA announced that microfossils found in an Antarctic meteorite could have come from Mars.[19] The meteorite, called Allen Hills 84001, was thought to contain frozen bacteria that was blasted from the Martian surface over 16 million years ago, when the Solar System was forming, and landed on Earth about 13,000 years ago. Debate continued on for years, with further investigations of the meteorite resulting in controversy on whether the lifelike patterns seen were really evidence of life. Further experiments and analysis with new technology has resulted in NASA scientists' current thinking that the bacteria could have survived the blast (rather than being caused by it) and was then transported to Earth. The material surrounding the Martian bacteria showed evidence of interaction with water on Mars more than 3.5 billion years ago.[20]

Why is discovering the evidence of life on Mars so important? We could say that life on any other planet, moon, or heavenly body changes everything about our perception of life being unique to Earth. This is similar to the shift in understanding from the scientific discovery that Earth is not at the center of the Solar System or that our Solar System is also not unique. If we are not the only planet or heavenly body with life, many more questions arise about the universe and its origins and our whole perception of our existence and even our notion of spirituality is thrown into doubt. These issues drive civilizations to find answers through exploration and scientific discovery.

How Much Will It Cost to Go to Mars and Why It Is Worth It

The debate about whether the United States should support manned space exploration today and in the near future compares predicted positive outcomes of colonizing outer space against risk and cost. Large sums of US federal monies have gone to defense and supporting wars in the past decade. Prohibitively high costs will be the most obvious deterrent. An expert panel

[19] Bruckner, Adam P. Humans To mars—why and how on earth are we going to do it? The Seattle Times. 17 Aug 1997.

[20] Leach, Ben. Nov 26, 2009. Bacteria from Mars found inside ancient meteorite. The Telegraph. [Internet] [cited 2015 April 16]. Available from: http://www.telegraph.co.uk/news/science/science-news/6660045/Bacteria-from-Mars-found-inside-ancient-meteorite.html

in 2014 estimated the total budget for a manned mission to Mars to be in the range of $80–$100 billion.[21] Most of the expenditure involves the development of NASA's Orion spacecraft and the new heavy-lift rocket, the Space Launch System (SLS). The SLS is a solid rocket booster rocket similar to the Space Shuttle Launch System but improved with the latest technology and the capability to carry humans and equipment to Mars and beyond.

After delays in the development of the SLS, it is estimated that the first test launch won't be before the end of 2018.[22] Continued on-schedule development depends on consistent NASA funding.

Considering the high cost in terms of NASA's budget and development uncertainties, scientists and space exploration advocates compare costs with benefits. Here are several reasons given for NASA to pursue manned missions to Mars.

1. America needs to embark on another great adventure and focus on a positive experience that boosts national pride.
2. Knowledge of the planet and the search for extraterrestrial life is an important goal. Robotic rovers can transmit only certain kinds of information. The search for fossils, and the ability to travel long distances and do heavy work are all beyond the scope of a rover. These activities if successful will further our knowledge of Earth's place in the universe and hold the key to its formation.
3. Technological and medical discoveries and innovation have been some of the primary benefits of human space exploration. The colonization of another planet, moon, or asteroid will force new methods of sustainable living, efficient transportation systems, and medical treatments.
4. Educational benefit that can result from a mission to Mars is a renewed interested in math and science for children and young adults who are inspired by a voyage to another planet.
5. Providing a possible habitat that may be required by humans in the future is another benefit. If we can learn to live on Mars, it might provide an option for life if there is a disaster on Earth or at the very least, a steppingstone to further exploration.

A manned mission to Mars is a very risky endeavor. Some of the risks associated with manned Mars missions include the following:

[21] Kaufman, Marc. 23 April 2014. A mars mission for budget travelers. National Geographic. [Internet] [cited 2015 May 25]. Available from: http://news.nationalgeographic.com/news/2014/04/140422-mars-mission-manned-cost-science-space/

[22] Morring, Jr., Frank. 24 Nov 2014. Work in progress. Aviation Week & Space Technology. 176:41:44–45.

1. Mars is about 150 times further away from Earth than the Moon is. The trip to Mars will last several months, making it the longest manned space mission in history. Astronauts have stayed on the International Space Station (ISS) for about that length of time, but the difference in traveling to Mars is that the astronauts are moving further and further away from Earth over time, making any chance of rescue nearly impossible. Issues that can arise over that time are human illness, mechanical failures, fire, micrometeorite collisions, exposure to deadly radiation and solar particle events, guidance and navigation errors, etc.
2. The mission has to support humans for the duration of the flight, including sufficient food, water, and oxygen. All of these sustainable systems have to work correctly to maintain life support.
3. If the mission is two-way, a vehicle launched from the surface of Mars has to be included and must perform satisfactorily, or the crew will not be able to return to Earth.

Both manned and unmanned space missions have suffered failure and tragedies over the years. Everyone accepts a certain amount of risk and uncertainty in space travel. The debate is about how much risk is acceptable for a mission to Mars to be launched. Manned scientific missions are very expensive and some think the funding would be better spent on more robotic missions or other types of scientific research.

The two manned Mars missions that have the most attention are the NASA Mars mission and the Mars One mission. Let's explore the similarities and differences in these expeditions.

NASA Mars 2020 Mission

NASA has invested in Mars exploration for decades, starting with a flyby and the first close-up picture of Mars in 1965. An impressive series of robotic rovers followed, transmitting detailed images of the landscapes and soil. Among the discoveries, one of the most important was the possible presence of liquid water, a primary necessity of life. Several successful missions set the stage for the ultimate journey for humans to Mars in the 2030s. The more recent and future missions are shown in the timeline in Fig. 3.3, with the science strategies outlined in Fig. 3.4.

The human mission to Mars will involve taking a trajectory that will take advantage of the orbits between Earth and Mars. It is preferable to choose a path that will use the least amount of fuel, which translates into the least

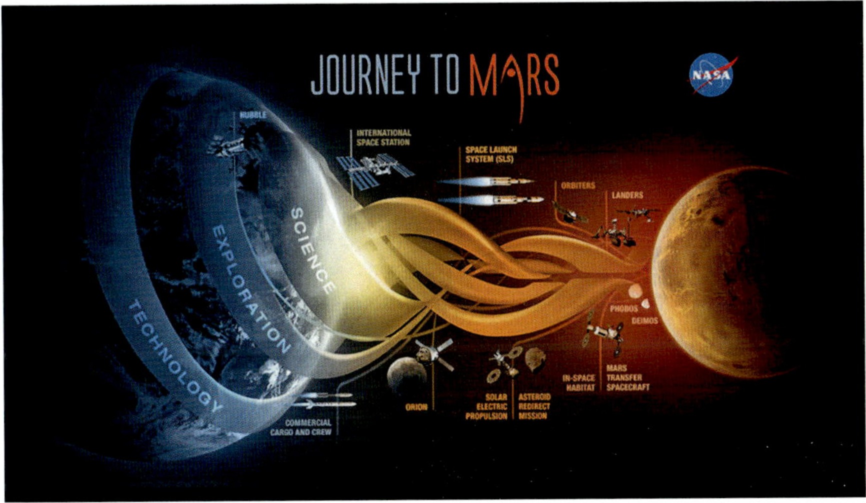

Fig. 3.3 NASA's journey to Mars (Image courtesy of NASA)

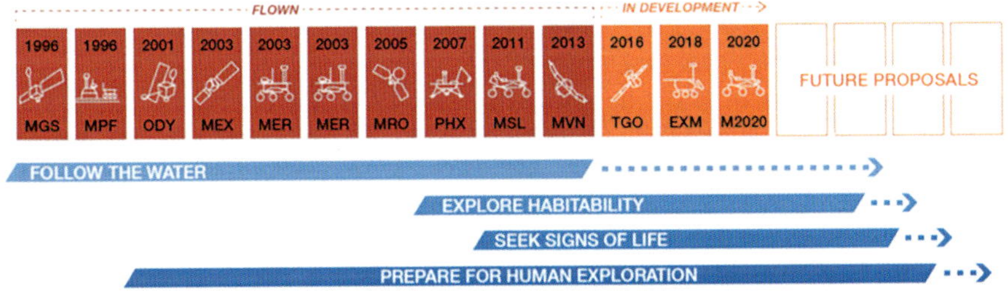

Fig. 3.4 Evolving science strategies for Mars exploration (Image courtesy of NASA)

amount of energy. In interplanetary travel, this is not a straight line. The spacecraft on the launch pad is already in Earth's solar orbit. Rockets are launched into the direction of this orbit and assisted by the rotation of Earth. The journey has to begin at exactly the right time in order to arrive at the correct point in space for their target, which is approaching. This time interval is called the launch window. The trajectory path is usually a partial orbit around the Sun. The spacecraft is launched into a transfer orbit moving around the Sun and toward the orbit of Mars. This is called a Hohmann transfer orbit (Fig. 3.5). Once the initial acceleration of the spacecraft increases the energy to the transfer orbit, essentially the vehicle can coast at that velocity until it reaches the target, which is an orbit around Mars. Timing is everything.

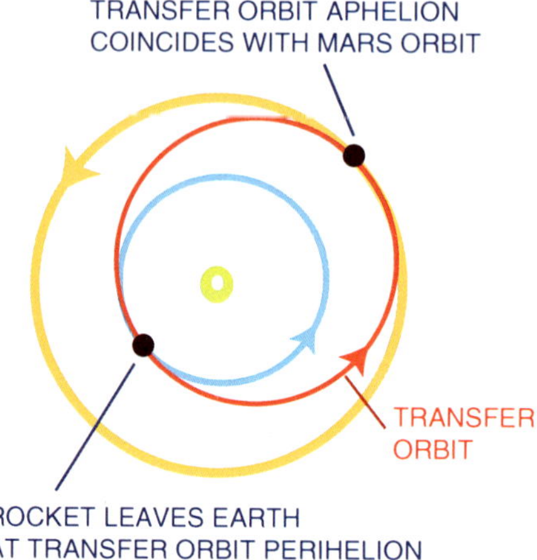

Fig. 3.5 Earth to Mars using a Hohmann transfer orbit (Image courtesy of NASA)

Arriving at the Mars orbit at exactly the right time and place require precise guidance and navigation. Both objects are moving at a high velocity. Once the Mars orbit is reached, the spacecraft will slow down by using a retrograde rocket or other method. In order to land, the vehicle has to decelerate more to enable it to land safely on the surface.[23]

The NASA Mars mission depends on continued government funding and other investments, including international partners. The Mars Exploration Program includes the development of a new spacecraft and preparation of an appropriate landing area by a newly designed robotic rover. The rover would use a guided approach, descent and landing, including the use of a parachute. The final landing involves what is referred to as a "skycrane maneuver" to lower the rover to the surface.

The proposed Mars 2020 rover mission would not only continue to look for life but also test key challenges for human expeditions using lessons learned from Curiosity. The rover will have the latest technology for analyzing, communicating, taking images, grappling for objects, etc. It will study rocks and soil, seeking signs of ancient life forms, monitor the Martian atmosphere and dust levels, and test the ability to extract oxygen from the primarily carbon dioxide atmosphere for future human missions.[24]

The astronauts will leave Earth aboard the Orion spacecraft atop the very powerful heavy-lift SLS rocket. Both are new designs (the SLS will have some

[23] Jet Propulsion Laboratory. Basics of space flight. Chapter 4. Interplanetary trajectories. [Internet] [cited 2015 April 28]. Available from: http://www2.jpl.nasa.gov/basics/bsf4-1.php

[24] Cole, W. UH trio picked for mars mission support. Honolulu Star—Advertiser. 22 Aug 2014.

existing shuttle technology) and are being tested. Orion is the first spacecraft built for space exploration since the Apollo missions and looks very much like the Apollo crew capsule. Orion will provide a safe environment for deep space travel, missions that extend much farther than previous manned missions.

One of Orion's missions in the 2020s will be to send astronauts to explore a large asteroid near Earth and collect samples using tools and techniques to be used for the manned mission to Mars (see below). Orion completed its first test in space on December 5, 2014, launched atop a Delta IV heavy lifting rocket, a proven multi-stage rocket with many commercial and military uses.

The Orion crew module was launched from Cape Canaveral, Florida, and landed less that 5 hours later southwest of San Diego. During the flight, Orion passed through the Van Allen Belt twice, where it was exposed to high amounts of radiation, heated up to 2200 °C on re-entry, and reached speeds of 32,000 km/hour. The successful testing of the Orion has been critical to the success of future manned Mars missions. Future test flights will include the integrated Orion and SLS system.[25]

The Asteroid Redirect mission (ARM) aims to capture a boulder from the surface of a near-Earth asteroid and place it into a stable orbit around the Moon. Near-Earth objects are defined as being within 1.3 AU of the Sun, one AU being the mean distance between Earth and the Sun, which is approximately 150 million km or 93 million miles. Asteroids that orbit the Sun are fragmented remnants of the Solar System's formation. Studying samples could provide new scientific information about the beginnings of life on Earth. The identification of a specific asteroid will not be announced before 2019. The possible asteroids will be chosen based on certain characteristics that would allow them to be redirected to an orbit near the Moon. NASA's Near-Earth Object Observation Program has cataloged more than 1000 new near-Earth asteroids. This program is part of the Asteroid Initiative of 2013, which includes ARM as well as an Asteroid Grand Challenge to locate potentially dangerous asteroids through partnerships with other organizations. Many more asteroids will be discovered over the next few years, and NASA will determine which ones have an optimal velocity, orbit, size, and spin to be the target asteroid. The four possible candidates identified so far are Itokawa, Bennu, 2008 EV5, and 1999 JU3. EV5 is an ideal candidate, with a diameter of approximately 400 m and with surface boulders as small as 7.5 m and smaller. Its last flyby to Earth was in 2008, and it won't be back until at least

[25] Britto, D. Space exploration in 2014: A year of achievements and discoveries. Daily News & Analysis. 30 Dec 2014.

Fig. 3.6 NASA Asteroid Redirect Mission—grabbing a boulder off of a near-Earth asteroid (Image courtesy of NASA)

2023. An ARM rocket launched in 2020 could catch up to the asteroid and perform the mission (Fig. 3.6).[26]

On March 25, 2015, NASA announced the method chosen for the robotic phase of ARM. A spacecraft will retrieve a boulder up to 4 m across from the surface of the selected larger asteroid and then return it to an orbit around the Moon. Important Mars mission technology, such as landing on the asteroid, and grabbing onto and moving the boulder, will be tested throughout the mission. In addition, planetary defense methods may be demonstrated to ward off future collision threats with Earth.

There is skepticism about ARM and how it relates to putting an astronaut on Mars. Arguments in favor of the mission include the implementation of new technologies that will directly relate to deep space tasks, such as advanced solar electric propulsion, new guidance and navigation techniques, and a staging point for which astronauts could enter deep space. The manned portion of ARM won't occur until 2025 and will use the Orion crew module atop the SLS. Orion would rendezvous with the boulder and perform crew EVAs to inspect the surface (Fig. 3.7).[27]

NASA's SLS is a composite of the space shuttle and the cancelled Constellation program (a NASA manned spaceflight program from 2005

[26] NASA unveils plans to pluck rock off asteroid. Telegraph—Herald. 26 Mar 2015.

[27] Gates, Michele. June–July 2015. The asteroid redirect mission and sustainable human exploration. Acta Astronautica. 111:29–36.

Fig. 3.7 NASA's Orion spacecraft after its successful flight test in December, 2014. (Image courtesy of NASA)

to 2009 that planned on travel to the International Space Station (ISS), the Moon, and Mars as the final goal). There are fifteen RS-25s, the space shuttle main engines, available for use. These engines would have to be refurbished for future missions, but the fact that this engine had a flawless record during the space shuttle program is one reason to use this technology and the available inventory. The SLS will also borrow certain features from the Ares I rocket (built for the Constellation program), including a new digital engine controller.

The SLS will be more powerful than the Saturn V multistage rocket that took astronauts to the Moon. As awesome and successful as this rocket is, its budget is astronomical as well. It is estimated that NASA will spend over $12 billion in development prior to a test launch in 2018. Costs could increase, and it is estimated that the project is already short on funding. The SLS could also be used for other missions, such as transporting American astronauts and goods to the ISS. In its first stage, the payload capability will be 77 tons and it will stand 321 feet, almost as tall as the Saturn V but with 10 % more thrust. As the rocket is further developed, it will stand on the launch platform at 40 feet taller than the Saturn V, with 20 % more thrust, which will enable it to

Fig. 3.8 Artist rendering of NASA's SLS rocket in a 70 metric-ton configuration as it lifts off into space (Image courtesy of NASA/MSFC)

deliver 143 tons to Earth orbit or a manned crew to Mars. It will be the most powerful rocket ever built.[28]

Much like the Saturn V Moon rocket, the Orion capsule for manned flight will be situated at the top of the stack. Below the capsule, the SLS will look very much like the space shuttle, with two solid rocket boosters attached to the sides. The first stage of the rocket (which NASA refers to as the core) is a stretched version of the shuttle's external tank with the same diameter, allowing the shuttle tools to be used. The bottom of the core will hold four shuttle engines (SSMEs). Finally, a single upper stage will boost Orion into deep space. Modern manufacturing techniques, such as 3D printing, will be used to build state-of-the-art equipment for the Orion capsule (Fig. 3.8).

The SLS will be very expensive with a lot of unknowns. Even though annual costs are much less than the space shuttle program, the shuttle made multiple flights (133 in over three decades), whereas the SLS will only launch once every 2 years. The Government Accountability Office estimates the total cost through 2020 at $22 billion and warns that the initial launch could be delayed without additional funding. However, the SLS is seen by NASA as a rocket that satisfies multiple needs and fills the gap post-shuttle for a variety of missions. It is possible that private sector partnerships can help fund the program.

[28] Betz, Eric. Jan 2015. Lofty goals, loftier budgets for human spaceflight. Astronomy. 43:1:16.

Fig. 3.9 Artist rendering of NASA's Low-Density Supersonic Decelerator (LDSD) (Image courtesy of NASA)

Landing on Mars has previously relied on one or multiple large parachutes to slow a vehicle's speed to prevent a crash landing. However, the Mars atmosphere is much less thick or dense than Earth's and doesn't slow the spacecraft down as much. The Curiosity lander carried a large mass of 900 kg (1 ton), requiring a giant parachute and descent rockets to slow the craft to a safe landing. Some Mars missions could double the mass of the payload, and the required parachute size to slow the vehicle down becomes too large (60 m in diameter). To assist in using atmospheric drag for the entry, NASA's Jet Propulsion Laboratory is developing a Low-Density Supersonic Decelerator (LDSD) Technology Demonstration mission, which will be tested full-scale in Earth's stratosphere. Currently, three types of devices are being built. The Supersonic Inflatable Aerodynamic Decelerator (SIAD) is a durable inflatable cone-shaped object that envelops the spaceship, increasing its surface area and its associated drag, thus slowing the craft down from Mach 3.5 or greater to 2. A 30.5-m diameter parachute is also being developed to slow down the vehicle further to subsonic speeds. These devices will assist with both unmanned and manned mission to the Martian surface (Fig. 3.9).[29]

[29] Steitz, David. May 2014. Low-density supersonic decelerator. NASA Press Kit. [Internet] [cited 2015 June 06]. Available from: http://www.jpl.nasa.gov/news/press_kits/ldsd.pdf

The specific plans for the SLS missions are still being solidified. The first mission was an unmanned test, sending the Orion capsule around the Moon and back to Earth. It was successfully completed in December 2014. The second mission plans to send astronauts to explore an asteroid orbiting the Moon (see the earlier description of the Asteroid Redirect Mission). Finally, the manned Orion trip to Mars will take place around 2030 and will last 18–24 months. It is possible that the astronauts' first mission will study the Martian surface from orbit.

There are many critics of the NASA Mars mission. Former NASA deputy administrator Lori Garver argues that by the time the SLS goes to Mars, the equipment will be "50-year-old technology." There are several new companies worldwide that are developing rockets for future manned and unmanned commercial missions, providing other ways of exploring deep space. One example is SpaceX. Its *Falcon 9* rocket has launched cargo to the International Space Station and delivered satellites to higher orbits. SpaceX is planning on building a heavy-lift rocket using advanced propulsion technology (liquid methane as fuel rather than the standard hydrogen or rocket-grade kerosene). In addition, its first stage booster is designed to vertically land on a floating platform in the ocean. These advances in rocket development have the support of many scientists and businesses. Another company, Orbital Sciences Corporation, is proposing future applications for its Cygnus spacecraft as a crew support vehicle to augment the capabilities and life support of Orion.[30]

One could argue that there is room for everyone. However, NASA has a limited budget, and many feel that the Mars mission is usurping the budget from other important endeavors planned for the future, such as a robotic mission to Europa. On the other hand, NASA has not extended beyond low Earth orbit with manned vehicles for decades and needs a bigger rocket to travel into deep space and maintain a leadership role in human spaceflight. The traditional methods of building a rocket require that the rocket be big and very expensive, and it is unclear if NASA will have the continued budget to support the SLS.

The Mars One Mission

The Mars One endeavor was designed by a Dutch space entrepreneur, Bas Lansdorp, the co-founder and CEO of the Mars One nonprofit organization. The plan, announced in 2012, has drawn a lot of interest and scrutiny because

[30] Chiles, James R. Oct/Nov 2014. Bigger than saturn, bound for deep space. Air & Space Magazine. 29:5:20–27.

it is not only ambitious but also controversial—start sending humans on one-way trips to Mars by 2025. The purpose is to colonize the planet by sending groups of manned missions staggered over time to build the first permanent human settlement on the surface of Mars. Mars One is taking on a similar role to NASA. The organization is coordinating technical efforts, training crews, and developing mission profiles. The actual rockets, living modules, landing rockets, etc., will be contracted out to developers.

The Mars One program is funded using a complex model that includes both a not-for-profit foundation and a for-profit corporation. The financial model is a unique way to fund such an expensive mission with the capability to capture the imagination of millions. Mars One relies on its exposure to millions through media, advertising the mission as the greatest human exploration. The organization receives donations from countries all over the world, from people who want to be part of this great adventure and from those who want to invest in the future. The non-profit foundation (Dutch name of "Stichting Mars One") will train the crews and own the colony on Mars. The for-profit corporation will finance the mission through investments, including a stock exchange listing and intellectual and media property rights. The mission can be accomplished using existing technology, but more efficient ways of living using fewer resources and innovation will foster new sustainable methods and inventions. This is the plan, but many naysayers are doubting the organization's ability to raise a sufficient amount of money in time and complete the necessary technical requirements prior to the estimated launch.

An engineering panel at MIT recently questioned the technical feasibility of the proposed mission. The results, published in October 2014, identified new technologies that will be needed to sustain life on Mars. Even though evidence of ice was discovered by the Mars Phoenix lander and Curiosity, the ability to extract the water from the soil has not been developed to such a point that it can be used for this application. In addition, Mars One proposed that all food be produced by the colony, which the panel states will produce unsafe levels of oxygen, requiring an oxygen removal method be implemented. Other estimations of budget and equipment requirements of the Mars One missions were considered to be overly optimistic, such as the number of Falcon heavy-lift rockets required to send up initial supplies. Mars One estimates six rockets would send up sufficient supplies, but the MIT team estimated that 15 rockets would be required. The associated costs would go up accordingly. Add to this the complexity of putting together this plan that relies on each step being tested prior to use. Although Mars One estimates that each launch will cost $6 billion, the MIT study predicts that the cost will be much higher.

The NASA two-way trip to Mars, which is supposed to be a cost-saving mission, is estimated to cost about $100 billion.[31]

Olivier de Weck, an MIT professor of aeronautics and astronautics and engineering systems, said in the *MIT News*, "We're not saying, black and white, Mars One is infeasible." Nonetheless, they think that "it's not really feasible under the assumptions they've made," he added. "We're pointing to technologies that could be helpful to invest in with high priority, to move them along the feasibility path."[32]

Prior to humans leaving on their mission, living modules will be transported to Mars. Several missions approximately 2 years apart are scheduled to prepare the colony in steps, starting with the use of an intelligent rover that will locate and prepare a suitable area for the modules and solar panels. Six cargo units are currently scheduled to be launched in 2024. They will be transported and set up by the rover, including connecting to a Life Support System (LSS), deploying solar panels and preparing the colony for human life. There are a lot of environmental factors and long term dangers that have to be addressed for a permanent colony. Some of these issues include dangerous levels of radiation, extreme temperatures, and an inhospitable atmosphere. The LSS will be connected to the living units to provide water, air, and electricity. Water has to be extracted from ice in the Martian soil and then condensed back to a liquid state and stored to be used for breathable air. The rover will deposit Martian soil as radiation shielding for the inflatable sections of the habitat.

The frontrunner for the Mars One heavy-lift launch vehicle is the new SpaceX Falcon Heavy (FH) rocket, which looks very much like the SLS with boosters strapped to the sides. (The role of SpaceX and its rocket development in the future of space exploration was discussed in more depth in Chap. 1.) SpaceX has been developing this powerful rocket in order to fill a need for both government and commercial space missions. This rocket is set for its first launch in November 2016.[33] The maiden flight could include a satellite payload, due to interest by many companies to participate in this historic mission. In 2014, SpaceX signed a lease agreement with NASA for the commercial use and operation of Launch Complex 39A, originally used for moon launches (Fig. 3.10).

---

[31] Do, Sydney, Ho, Koki, Schreiner, Samuel, Owens, Andrew, de Weck, Olivier. An independent assessment of the technical feasibility of the mars one mission plan. 65th International Astronautical Congress, Toronto, Canada. Sep 29–Oct 3, 2014.

[32] Chu, Jennifer. Mars one (and done?). MIT News. 14 Oct 2014.

[33] Gough, Evan. SpaceX maiden Falcon heavy launch may carry satellite in November. Universetoday.com. 09 May 2016. [Internet] [cited 2016 June 10]. Available from: http://www.universetoday.com/128815/spacex-maiden-falcon-heavy-launch-may-carry-satellite-in-november/

Fig. 3.10 Gwynne Shotwell, President and CEO of SpaceX announcing a lease agreement with NASA for the use and operation of Launch Complex 39A, originally built for Apollo/Saturn V rockets. NASA Administrator Charlie Bolden on left and Kennedy Space Center Director Bob Cabana on the right. (Image courtesy of NASA)

One of Falcon's capabilities will be to deliver satellites to a geosynchronous orbit, which is a high Earth orbit (22,236 miles, or 35,786 km above Earth[34]) that matches satellite speeds to Earth's rotation, allowing for stationary surveillance and communication networks. The geosynchronous orbit is occupied primarily by communication and weather satellites. The orbit classifications and altitudes are shown in Fig. 3.11. The last rocket to have the capability to deliver more payload than the SpaceX Falcon to Earth orbit was the Saturn V Moon rocket, last flown in 1973 and decommissioned after the Apollo program.

How does the SpaceX FH lift rocket compare with the SLS? Let's compare the two heavy-lift launch vehicles being developed separately by NASA and SpaceX by evaluating the progress in both programs and the projection of what will be flying in the 2018–2019 timeframe.

[34] Riebeek, Holli. Catalog of Earth satellite orbits. Earthobservatory.nasa.gov. 04 Sep 2009. [Internet] [cited 2016 June 10]. Available from: http://earthobservatory.nasa.gov/Features/OrbitsCatalog/

Fig. 3.11 Orbit classifications and their associated altitudes. (Image courtesy of NASA)

The SLS and the FH look very similar, with boosters strapped to the side around a central core that has a second stage on top and the payload above that. The design approach similarities end there.

The SLS is designed to achieve certain heavy lifting goals (70 MT, then 105 MT, and finally 130 MT) and to reuse the existing technology available from the space shuttle and Constellation programs (completion of the International Space Station and the return to Moon mission). There is also a modification of proven Delta IV upper stage, a rocket engine that burns liquid oxygen and hydrogen and provides a maximum terminal velocity. The rocket design takes a traditional approach using solid rocket boosters to yield greater thrust for a given size and weight as well as reusing the established space shuttle main engines, also using liquid oxygen and hydrogen for fuel.

On the surface, it might seem like the SLS design is a cobbled connection of used parts. However, this approach makes the rocket more affordable, time tested, and more easily manufactured.

The FH rocket is based on the technology of the Falcon 1 and Falcon 9 rockets. The central core, boosters, and upper stage appear to be either unmodified or slightly modified from the Falcon 9. The proposed engines (a total of 27) are the Merlin IDs that use liquid oxygen for fuel, and the RP-1, using liquid oxygen and hydrogen. These engines are optimized for minimum cost to low Earth orbit with the option to transfer to a geosynchronous orbit. The engines provide less thrust than the SLS engines but run cooler, with less wear on the parts and greater reusability. Emphasis was also placed on minimizing the cost of manufacturing and operations.

A cost comparison of capabilities to low Earth orbit is quite dramatic. The payload capability of the first SLS design is 70 MT, costing $600 M, yielding a cost per MT of $8.6 M/MT. Payload capability of the FH Heavy is 53 MT, costing $158 M, yielding a cost per MT of $3.0 M/MT. Clearly, the proposed FH Heavy design is more cost effective to low Earth orbit. However, the SLS will take the advantage for launching further out in space. SpaceX remains as a cost effective solution for business space applications. The SLS is built for heavy lifting into deep space.

The risk factors are similar for both designs. Both are using some proven technologies with some modifications. SpaceX has the edge in terms of consistent testing and modifying its designs. However, when we move to large payloads being transported into deep space, the SLS begins to have the edge. SpaceX plans to develop its most powerful engine, a methane-liquid oxygen engine delivering 500,000 pounds-force compared to the 150,000 pounds-force for the Merlin 1D. This engine would be responsible for the heavy-lift capabilities to Mars and beyond. Both rockets will most likely be developed and used for perhaps different types of missions. Whether we go to Mars or back to the Moon first, it is possible that both rockets will play a part.[35]

What are the human aspects of the colonization? When Mars One announced its plan and put out its initial call for a crew to set up the colony, the selection committee received over 200,000 responses. The committee is now down to 100 finalists, 50 men and 50 women from all over the world with a variety of backgrounds and life situations. The concept of this journey is seen as being very risky and controversial because of the untested new technologies employed and the one-way ticket advertised as a unique opportunity to colonize a planet. One of the finalists is featured below.

The colonization plan is ambitious, sending groups of four humans (two men and two women) at a time on the journey to Mars, starting in 2025. The finalists were originally to begin training in 2015, but announcements were made that further rounds of selection will be necessary and will not take place until the end of 2016–2017. Once the final number is reduced to six teams of four, training will begin. The training will entail spending long periods of time in remote locations, where the applicants will practice being self-sufficient in all aspects of their lives. They will endure physical and psychological hardships similar to the isolation of a permanent settlement on Mars. The final 24 men and women chosen will have trained for up to 10 years.

Sonia Van Meter, mid-30s in age, of Austin, Texas, is a managing director of a political consulting firm, wife, stepmother of 2, and one of the 100 finalists selected for the one-way trip to Mars to set up a human settlement.[36] Sonja is a woman who acted on what many of us can only dream about expe-

[35] Skran, Dale L. 27 April 2015. Battle of the Collossi: SLS vs falcon heavy. The Space Review. [Internet] [cited 2015 May 27]. Available from: http://www.thespacereview.com/article/2737/1

[36] Texas Monthly Staff. Honey, I want to move to Mars. 2 Apr 2014. [Internet] [cited 2016 Mar 26]. Available from:

http://www.texasmonthly.com/the-culture/honey-i-want-to-move-to-mars/

riencing in our lives. Historically, humankind has pursued a number of great adventures without knowing the outcome; mariners set out on an open sea to discover a new world; cosmonauts and astronauts became the first humans launched into space; mountain climbers pursued the tallest peaks on Earth. Most of us either long for such an adventure or can at least appreciate those who pursue such risky achievements. Sonia hopes to be chosen as one of the final 24 humans to set out to colonize Mars. Since Sonia's story has been publicized, there has been a large public outcry with her decision to leave her husband and children forever if she is chosen to continue on a path that will allow her to pursue a dream alone.

Laurel Kaye at only 21 years old is dreaming of being on the surface of Mars by the time she is in her mid-thirties. Not that long ago, women were restricted from being pilots or astronauts, and now she is hoping to be part of a Mars One crew that will live on Mars. Laurel embodies youthful optimism and an adventurous spirit: "Whether it succeeds or fails, we stand to learn so much from something like this. At some point, you just have to be the first person to do something. Otherwise, nothing is going to happen."

Laurel is a physics graduate from Duke University with a promising future ahead, whether it includes Mars or not.[37]

Colonization could include populating the planet, although that is not a mission objective. It is possible and probably inevitable that teams with equal numbers of men and women will eventually produce children on Mars. One of the five British candidates in the final 100 candidates is interested in giving birth on Mars. Maggie Lieu, an astrophysics Ph.D. student, would be interested in being the first mother on Mars, giving birth to the first Martian. Of course, giving birth in a low-gravity environment could be a challenge and has never been done. A further discussion of low-gravity effects on the body is in Chapter 5.[38]

Recently, one of the finalists, exposed what he thinks is a scandal in the Mars One selection process. The former NASA researcher says that applicants were not put through extensive interviews or testing, and as a finalist, he never met anyone from Mars One in person. In addition, he says that the process is extremely commercial, leaning heavily on the finalists' ability to donate

[37] Duke University Dept of Physics Website. [Internet] [cited 2016 Mar 26]. Available from: https://www.phy.duke.edu/news/physics-undergrad-laurel-kaye-finalist-mars-one-colony

[38] Rowley, Tom. The Telegraph. British astrophysics students wants to have 'first Martian baby'. 17 Feb 2015. [Internet] [cited 2016 Mar 26]. Available from: http://www.telegraph.co.uk/news/newstopics/howaboutthat/11417231/British-astrophysics-student-wants-to-have-first-Martian-baby.html

money to the program. Time will tell where the truth is in this process and if the program is flawed using this procedure.[39]

Final Thoughts on Humans Going to Mars

It is difficult at this time to compare the two endeavors designed to send humans to Mars. It might seem that NASA is almost too methodical, following a painstakingly slow process to develop a system sufficiently safe to launch humans into deep space. Mars One seems to take a riskier approach along with a questionable process of sending humans on a one-way trip to colonize another planet. At the moment, both organizations are going forward with their plans (although there hasn't been any updates from Mars One recently).

A trip to Mars will rekindle the enthusiasm for space exploration in a similar way to the summer of 1969. Exploring and colonizing another planet is the next big step for manned space flight. The goal is not just a rebirth of a sense of adventure in the public eye but also to bring us new scientific information about Mars that will give us an understanding of the past and future of Earth.

As Buzz Aldrin has said: "Just as Mars—a desert planet—gives us insights into global climate change on Earth, the promise awaits for bringing back to life portions of the Red Planet through the application of Earth Science to its similar chemistry, possibly reawakening its life-bearing potential."[40]

Critical steps in the process to put American astronauts on Mars are planned for the next couple of years. Safety, of course, is a primary factor. We are all anxious for the Mars missions, but we also know that safety is a result of extensive planning for backup systems and a protection of life. Nonetheless, many of us can't wait for the day when we walk on another planet.

[39] Bowerman, M. Want to live (and die) on mars? USA Today. 18 Feb 2015.

[40] Aldrin, Buzz. Commentary: let's aim for Mars. CNN.com. 23 June 2009. [Internet] [cited 2016 Feb 12]. Available from: http://www.cnn.com/2009/TECH/space/06/23/aldrin.mars/

4

Why Not Go Back to the Moon?

Keywords George W.S. Abbey • Asteroid Redirect Mission • Astrobotic Technology Inc. • Charles Bolden • George W. Bush • Buzz Aldrin • Charles Bolden • China • Constellation Moon • George Abbey • International Space Station • Lunar CATALYST Initiative • Masten Space Systems Inc. • Moon Express Inc. • National Research Council (NRC) • Space Exploration Initiative (SEI) • Space Station Freedom • Vacuum

> "Well, I don't think we should go to the Moon. I think we maybe should send some politicians up there."
>
> –Ron Paul, 2012

Going back to the Moon doesn't seem as interesting as going to Mars. Humans have never set foot on another planet, and the challenge and intrigue of setting foot on the Red Planet far exceeds returning to the Moon. NASA's proposed manned Mars mission includes a preparatory unmanned journey to the Moon. The scientific community feels that going back to the Moon and setting up habitats to develop and test techniques for sustainable living would improve our chances for success on Mars and is the logical and safer progression than going straight to Mars.

We have landed on the Moon several times but know very little about it—its origins, resources below the surface, etc. Arguments can be made to explore the celestial body nearest to us and establish a colony that can explore available resources and long-term survival challenges while still within a reasonable distance from Earth. Others feel that focusing on the Moon takes

L. Dawson, *The Politics and Perils of Space Exploration*,
Springer Praxis Books, DOI 10.1007/978-3-319-38813-7_4

away resources that could be used for developing Mars destination technology. There are experts and logical arguments on both sides of this issue.

Finally, there is the international standpoint, whether politically the challenge should be to return to the Moon or be the first to go to Mars. Definitely focusing on the Moon will delay the mission to Mars, and the thought of being second or third in this effort is not acceptable to many. This chapter explores some of these options and provides some of the scientific and political arguments on both sides.

Going Back to the Moon

Discussions about going to Mars usually include a discussion of previous manned missions to the Moon and the idea of using the Moon as a launching platform to deep space. The Moon is also thought to have resources that could be accessed and mined to build launch platforms. It also makes logical sense to launch rockets from a site with less gravity in order to save fuel and to provide an efficient mechanical servicing facility and international launch site.

One argument for returning to the Moon is to use it as a launching platform or as a possible human settlement in order to explore or colonize the Solar System. Rockets designed to be launched directly from Earth to Mars for example have to be powerful enough to overcome Earth's gravity first and then travel the long distance to Mars. This has been accomplished previously by using multistage rockets that provide enough thrust for lifting heavy payloads off of the surface of Earth. As each stage burns out, it drops off to propel the next stage faster into space because it is lighter than the first. If you start on the Moon, you only have one-sixth of Earth's gravity to escape, which makes the rocket development and ability for long-term flight easier. In addition, rocket design is targeted for certain operating conditions, including the atmosphere. Rockets that operate efficiently in outer space with no atmosphere, called a vacuum (which includes the surface of the Moon), are designed differently than those that have to push through Earth's dense atmosphere.

Mining on the Moon can also lead to the discovery and extraction of materials that could be used for rocket development. It is thought that below the lifeless surface of the Moon, there are minerals and rare elements of national importance. Rare Earth elements are found in Earth's crust and are used for modern electronics, health care, and national defense applications. They have unique properties that include greater energy efficiency, durability, and reduced weight. There are only seventeen elements (including scandium, lanthanum, and cerium) that fall in the rare Earth element category and are a valuable commodity among nations that mine them. China is the leader in this mining activity, followed by the United States, India, and Australia. China has blocked the export of these elements to Japan to prevent their applications in wind turbine, hybrid care, and defense. As these elements become rarer, there are increasing concerns about the countries that control these resources. It is evident that some of these materials are contained inside of the Moon's crust similar to Earth's crust. However, more sophisticated research would have to be conducted to see where the rare elements are concentrated.[1]

Being close to Earth, the Moon provides a somewhat familiar space environment to test new technologies to be used in a more ambitious trip to Mars. Current national space policy calls for manned missions to asteroids and Mars but not the Moon. Some organizations, such as the National Research Council (NRC), feel this is the wrong approach. In a comprehensive review of NASA's human spaceflight program, the NRC concluded that NASA's strategy will be unsuccessful in achieving a Mars landing in the near future. The 286-page NRC report, mandated by Congress and published in June 2014, reviewed research collected over an 18-month period of time. It states that if NASA stays on its current course with a consistently low budget, it will "invite failure, disillusionment, and the loss of the longstanding international perception that human spaceflight is something the United States does best." The report supports sending astronauts back to the Moon (Fig. 4.1).[2]

The Politics of Returning to the Moon

The primary reason that the United States and the Soviet Union embarked on the Moon race was to gain political power and prestige during an intense Cold War. Since the last Apollo mission (17) in 1972 (Fig. 4.2), several missions from other countries were sent to the Moon, but only Americans walked on the Moon. The Smart-1, launched by the European Space Agency in 2003 using solar-electric propulsion, employed sophisticated miniaturized instruments to analyze key chemical elements on the lunar surface until its mission ended in 2006. The Japan Aerospace Exploration Agency (JAXA) launched

[1] David, Leonard. 04 Oct 2010. Is mining rare minerals on the moon vital to national security? Space.com. [Internet] [cited 2016 Apr 02]. Available from: http://www.space.com/news/moon-mining-rare-elements-security-101004.html

[2] Achenbach, Joel. NASA strategy can't get humans to Mars, says National Research Council spaceflight report. The Washington Post. 05 June 2014.

Fig. 4.1 *Apollo 11* lunar module looking back toward Earth from the Moon, 1969. (Image courtesy of NASA)

Kaguya (Selene) in 2007 to investigate the lunar origin and develop technology for future space technology.[3]

In mid-1989, on the twentieth anniversary of the *Apollo 11* lunar landing, President George H. W. Bush announced plans for the Space Exploration Initiative (SEI). Plans included the construction of the Space Station Freedom, returning humans to the Moon, and future manned missions to Mars. The study following this announcement resulted in an estimated cost of $500 billion, which was mindboggling even spread over 20–30 years. It was unable to gain congressional support, and President Bush was unable to pull international partners together at this high a cost. The resulting recommendation was that NASA focus on space and Earth science, and manned spaceflight would be addressed project by project. The result was that human exploration beyond Earth orbit was dropped, and robotic exploration became the only affordable option for the next several years (Fig. 4.3).[4]

---

[3] Bilger, Burkhard. Apr 22, 2013. The martian chroniclers—a new era in planetary exploration. The New Yorker. 89:10:64–79.

[4] Dick, Steve. Summary of space exploration initiative. Nasa.gov. [Internet] [cited 2016 April 02]. Available from: http://history.nasa.gov/seisummary.htm

Fig. 4.2 Harrison Schmitt, Apollo 17 lunar module pilot uses a sampling scoop to retrieve materials. (Image courtesy of NASA)

Fig. 4.3 Space Exploration Initiative (SEI) lunar base concept (Image courtesy of NASA)

In 2003, President George W. Bush's administration boldly proposed that the United States go back to the Moon. The Constellation Moon landing plan never really got off the ground, due to lack of funding and changing political objectives. Some of the NASA budget was diverted to extend the life of the International Space Station to 2020. Additional portions of the budget focused on the development of spacecraft by private enterprise to ferry astronauts to the space station after the space shuttle was retired. When President Obama released his 2011 budget proposal, it was determined that the funds to return to the Moon were insufficient. NASA's future was left somewhat undefined, with only vague goals defined to send humans beyond Earth's orbit but with no specific timelines.[5]

The general consensus in early 2016 is that we've already visited the Moon (six Apollo missions with men walking on the surface of the Moon) and that it's time to travel beyond to discover new worlds. However, many experts think that we should return to the Moon to explore new territories and establish and test new technologies required for deep space travel and colonization. Both the Moon and Mars have similar gravities, atmospheres, and temperatures. A Moon colony could provide a scientific testing center using current technology to travel, land, set up living quarters, and test sustainable habitats with less risk than travel to Mars. Former NASA flight director George W. S. Abbey says, "You're not going to go to Mars before going back to the Moon. You need to establish a goal to go to the Moon and do that first and have a program laid out for an effective way to do it, but they're not doing that right now and I think that's really key to exploration."[6] In an age of budget constraints combined with instant gratification, public support is more excited about traveling to Mars than making preparations on the Moon for an upcoming mission to Mars. Given limited resources for the foreseeable future and the expense of manned missions, the choices of what and where to explore have to be prioritized.

Thus far, the Moon is seen as a more favorable destination by our international partners. Over 45 years after the United States won the space race to the Moon against the Soviet Union, when asked about NASA's plans to return to the Moon, Charles Bolden (then NASA Administrator and former astronaut) made it clear that NASA had no plans to have a human return to the Moon. "NASA will not take the lead on a human lunar mission," he said. "NASA is not going to the Moon with a human as a primary project probably in my lifetime.

---

[5] Mail FS. 02 Feb 2010. Obama scraps new moon mission. Daily Mail. [London (UK)]: 8.

[6] Burks, Robin. Tech Times. 7 Jan 2015. Should we go to the moon before mars? These Astronauts Think So. [Internet] [cited 2015 April 22]. Available from: http://www.techtimes.com/articles/24920/20150107/astronauts-think-we-should-go-to-the-moon-before-mars.htm

And the reason is, we can only do so many things." Instead, he said the focus would remain on human missions to asteroids and to Mars. "We intend to do that, and we think it can be done."[7]

There seems to be a distinct difference in goals between US space policy and other countries interested in setting up platforms and doing research on the Moon. Because international collaboration is important, it is useful to examine the disconnect in policies. Most industrialized nations have extensive lunar plans except for the United States. In 2010, President Obama canceled NASA's plans to return to the Moon in lieu of the Asteroid Redirect Mission, a precursor to travel to Mars.[8] This mission was discussed in detail earlier. The biggest reason for the focus on the Moon by nations other than the United States is cost. A manned mission to Mars requires a heavy-lift rocket with sophisticated technology to travel to the planet and return safely. This type of mission requires a long-term commitment and a tolerance for risk from an international economy that has been stagnant or growing slowly.

Currently, space agencies interested in sending humans to the Moon as early as 2025 include China, Japan, Europe, Russia, Iran, and a few private companies. Some plans are vague, but others are more specific, such as building bases, mining resources, and studying the surface in scientific detail. More nations are planning on sending robotic missions to the Moon. China put a robotic rover on the Moon in 2013, becoming the first nation to successfully make a soft landing there in over 40 years. Another milestone planned for China is to be the first country to land a probe on the dark side of the Moon by 2018.[9] Europe, Japan, India, Russia, North and South Korea and several private companies hope to follow in the near future with Moon explorations.

China is the country with the most ambitious space program, with plans to land humans on the Moon by 2036 in preparation for establishing a Moon colony. In order to make that a reality, China will need to develop a powerful rocket and more advanced technology. Advancing China's space program seems to be a priority for the Chinese government, and currently President Xi Jinping is pushing to establish China as a space power. Although insisting the program's goals are peaceful, the United States is cautious looking at some of China's demonstrated capabilities to prevent space-based assets from operating in a crisis. The

[7] Hedman, Eric. 04 Aug 2014. The Moon or Mars?. The Space Review. [Internet] [cited 2015 June 05]. Available from: http://www.thespacereview.com/article/2737/1

[8] Weitering, Hanneke. 04 Feb 2014. NASA opts out of new moon race. [Internet] [cited 2015 June 05]. Available from: http://scienceline.org/2015/02/nasa-opts-out-of-new-moon-race/

[9] Reuters.com technology staff. 15 Jan 2015. China to land probe on dark side of the moon in 2018: Xinhua. Reuters.com. [Internet] [cited 2016 June 16]. Available from: http://www.reuters.com/article/us-china-moon-science-idUSKCN0UT030

US Congress has banned NASA from cooperating with the Chinese for common space goals due to possible espionage, although China has indicated that they want to partner with the United States on space missions. China plans to launch a space station module around 2018 as a first step to a permanent operating station around 2022. The focus is primarily on the Moon, although future plans for China are to launch a probe to Mars in 2020.[10]

Wu Weiren, chief designer of China's missions to the Moon and Mars, told the BBC the following: "Our long-term goal is to explore, land, and settle [on the Moon]. We want a manned lunar landing to stay for longer periods and establish a research base." He also spoke on partnering with the United States: "We would like to cooperate with the US, especially for space and Moon exploration," he said. "We have urged the US many times to get rid of restrictions so scientists from both countries can work together on future exploration." Buzz Aldrin (retired Apollo astronaut) predicts that the United States will start working with China soon. "I think we will be organizing the other three—Russia, Europe, Japan—so that they will be cooperating and coming along soon after China, because we're helping all of them," he said.[11]

Time will tell whether NASA will need to alter its priorities about the Moon in order to succeed in its long-term vision of sending humans to Mars. For many scientists, it makes sense to develop a Moon base, test mining practices and set up commercial contracts and delivery methods for private enterprise in preparation for future activities and possible colonization in deep space. The politics of another country such as China setting up an extensive Moon base brings up Cold War feelings and political issues that will have to be addressed in refining international space policy and thinking about the importance of strategically exploring space.

All this being said, NASA has reinforced its commitment and readiness for the upcoming manned missions to Mars. In September 2015, Bolden stated that NASA's goal of getting astronauts to Mars by the 2030s is totally achievable. "We are farther down the path to sending humans to Mars than at any point in NASA's history," he said in a NASA headquarters meeting on manned Mars plans. "We have a lot of work to do to get humans to Mars, but we'll get there."[12]

---

[10] Harrington, Rebecca. Techinsider.io 21 Apr 2016. China plans to reach Mars by 202o and eventually build a moon base. [Internet] [cited 2016 June 16]. Available from: http://www.techinsider.io/china-plans-mars-moon-landings-2016-4

[11] Harrington, Rebecca. Techinsider.io 21 Apr 2016. China plans to reach Mars by 202o and eventually build a moon base. [Internet] [cited 2016 June 16]. Available from: http://www.techinsider.io/china-plans-mars-moon-landings-2016-4

[12] Space.com Staff. 5 Sep 2015. A manned mission to Mars is closer to reality than ever. Space.com. [Internet] [cited 2015 June 05]. Available from: http://www.space.com/30580-nasa-manned-mars-mission-reality.html

Commercial Interests in Returning to the Moon

There are commercial interests in returning to the Moon. Private enterprise is motivated by the commercial aspect of space travel rather than the science of exploration. Golden Spike is a private space company with interests in flying both manned and unmanned missions for international customers to the Moon by 2020.[13]

A program initiated by NASA early in 2014, called the Lunar CATALYST Initiative, seeks to partner with private companies to develop reliable and cost-effective delivery of payloads to the Moon's surface through the use of robotic landers. The initiative will provide companies who win the competition with scientific support, facilities, and technology rather than funding to support commercial activities. The three US company finalists selected are Astrobotic Technology Inc., Masten Space Systems Inc., and Moon Express Inc.[14]

Developing navigation and hazard avoidance for a self-landing, rocket-powered spacecraft on Earth is challenging, due to the need to test in the same operating conditions that the system would encounter in a planetary landing. Astrobotic and Masten companies collaborated on the technology that enabled a successful test flight without prior knowledge of exactly where the rocket would choose to land. Astrobotic's AAS scanned the landscape and selected a safe landing point. Masten's onboard flight system received input from the Astrobotic vision and navigation system, validated the input, and accepted the selection of a path to the touchdown point. The flexible architecture enables more choices for future landings on the Moon.[15]

Moon Express's commercial efforts are focusing on mining the lunar surface. In the near future, Moon Express will be exploring the Moon for the potential of mining precious materials needed on Earth, with the goal of returning lunar samples to Earth for science and commercial purposes.

Related to the Catalyst Initiative, astronaut and Moon walker Buzz Aldrin is supportive of the United States helping other nations travel in space: "Let's try doing something that doesn't compete with prestige-seeking nations sending their citizens to kick up dust on the Moon," Aldrin said during a

[13] Poppick, Laura. Space.com Staff Writer. 2 Jan 2014. Moon rovers planned for commercial lunar exploration project. [Internet] [cited 2015 April 23]. Available from: http://www.space.com/23946-golden-spike-private-moon-rover-designer.html

[14] NASA selects commercial space partners for collaborative partnerships. U.S. Fed News (USA). 25 Dec 2014.

[15] Moon Express, Inc. website. [Internet] [cited 2016 June 16]. Available from: http://www.moonexpress.com/missions.html

Google Hangout with Space.com in July 2014. "The United States should help other nations by placing robotic probes on the Moon that can be used to explore and aid other nations' lunar ambitions," Aldrin added.[16]

It remains to be seen what commercial efforts will succeed on the lunar surface but it is clear that the Moon will provide an important testing ground for a lot of new and sophisticated technologies with space applications.

[16] Kramer, Miriam. Space.com Staff Writer. 21 July 2014. The future of moon exploration, lunar colonies and humanity. [Internet] [cited 2015 April 26]. Available from: http://www.space.com/26584-future-of-moon-exploration.html

5

The Science and Dangers of Outer Space

Keywords Alzheimer's disease • Cardiovascular effects • Charles Bolden • G forces • Yuri Gagarin • Geosynchronous • *Gravity* • International Geophysical Year • Ionizing effects • ISS • James Van Allen • Japan's Aerospace Agency (JAXA) • Joint Space Operations Center (JspOC) • Kessler syndrome • Lockheed Martin • Dr. Jeffrey Lotz • Magnetic net—for space debris • Mazlan Othman • Meteor bumper • Microgravity • Muscle Atrophy • National Research Council (NRC) • Radiation • Shielding • Space Debris • Space Fence • Space Treaty of 1967 • Van Allen Belts • Werner Von Braun • Whipple shields • Dana Whalley • Fred Whipple

Robot: "It sounds like old Morse code."
Will Robinson: "What does it say?"
Robot: "Danger, Will Robinson, danger!"
Lost in Space, 1998

Lost in Space

Anyone who has watched a space-related science fiction movie knows that outer space is a dangerous and unforgiving place. In older films especially, humans exposed to the vacuum of space because of a malfunction or rip in their spacesuit experience a series of horrific events, including eyeballs bulging and popping out, the body swelling, blood boiling, the head exploding—eventually ending in blood squirting everywhere. Scientists now have a better understanding of what exposure to the airless void of space would mean to

L. Dawson, *The Politics and Perils of Space Exploration*,
Springer Praxis Books, DOI 10.1007/978-3-319-38813-7_5

a human body, based primarily on research on animals exposed to a vacuum and better scientific knowledge of the space environment.

Beyond Earth's protective atmosphere (approximately 62 miles, or 100 km, altitude above Earth), space cannot support human life. Some more modern science fiction movies such as "2001: A Space Odyssey" more accurately depict the result of exposure when astronaut Bowman goes outside of the spacecraft without a helmet to save a colleague. He is exposed to 14 seconds of the vacuum of space before he can close the airlock and pressurize the chamber. Previous research on chimpanzees in the 1960s by the air force demonstrated that someone could survive the 14 seconds, but it would likely be unpleasant and could result in a loss of consciousness. Survival might be possible for about 90 seconds of exposure.[1]

While exposed to no pressure, air would be expelled out of the body. Reduced pressure would also result in nitrogen dissolved in the blood to form bubbles known as the divers' condition called the "bends." Brain asphyxiation results because oxygen is not reaching the brain and is being expelled out of the blood, working in reverse of normal. Some of the effects would be similar to high altitude exposure such as impaired judgment and loss of vision. If exposure continues, unconsciousness, paralysis, convulsions and death would occur. Water vapor forming in soft tissue would cause the body to swell to as much as twice the normal volume. Heart rate slows, blood pressure drops and blood circulation stops within the first minute.[2]

Thus far, no accident has resulted in any human experiencing these symptoms or dying from exposure to the vacuum of space, but understanding this environment is important for research in space medicine. Because humans cannot survive outside of a spacesuit or spacecraft except for a few seconds, scientists and engineers understand and appreciate the importance of backup systems and providing complete protection of humans traveling long term through space. This chapter identifies the major issues that affect space exploration and the scientific challenges associated with addressing them.

---

[1] Bioastronautics Data Book. 1965. Aerospace Medicine. 36:9:890–&.

[2] Whitman, Justine. Aerospace Web. [Internet]. Aerospaceweb.org. c2012. Human exposure to the vacuum of space; Jan 28, 2007 [cited 2015 Aug 15]. Available from: http://www.aerospaceweb.org/question/atmosphere/q0291.shtml

The Space Environment

Humans on Earth live in a safe and supportive environment with an atmosphere that protects us from the extremes of outer space. We live in comfortable temperatures, acceptable pressures with breathable air and a gravitational force that keeps us attached to Earth's surface. As human beings step outside of Earth's protective atmosphere, all that disappears. Spacecraft with artificial environments are necessary to support human life. Spacesuits are required to step outside of the spacecraft. If either the spacesuit or spacecraft is compromised, a backup solution for providing life support is needed—immediately. Unsupported human life cannot exist in outer space.

Outer space is defined as the void that lies beyond Earth's atmosphere, starting at 62 miles (100 km) above Earth's surface. The void has no gas molecules, making it a vacuum. In other words, outer space is its own environment, with an atmospheric pressure of zero along with extreme temperatures. On the sunlit side of an object at Earth's distance from the Sun it would be over 120 °C (248 °F), while the shaded side would plunge to −100 °C (−148 °F). There are waves that flow freely through space, for example, radiation and light.

The gravitational pull of Earth on a spacecraft becomes less and less until microgravity or near-zero gravity is experienced. The prefix micro is added because the actual gravity measurement is not exactly zero. Astronauts in Earth orbit experience microgravity because of the complex balance of falling towards Earth due to the pull of Earth's gravity while traveling fast enough to match the curvature of Earth. Microgravity has a significant effect on bodily functions, both short and long term, and how daily activities are performed. If a human being were to be left unprotected in outer space, air in the lungs would rush out into the vacuum of space, skin would expand, bubbles would form in the bloodstream, and fragile tissues would rupture. Add deprivation of oxygen to the brain, and unconsciousness would occur in less than 15 seconds.[3]

Because of the extremes found in outer space, humans and their spacecraft have to be protected at all times. Spacecraft have to be constructed from materials that can withstand the harsh environment, including extreme temperatures and radiation, and manned missions have to have backup systems in place to protect the crews.

The focus in this chapter is on manned missions and the hazards on humans encountered in outer space and during deep space exploration.

[3] NASA Quest. NASA. [Internet]. Quest.nasa.gov. The outer space environment; Feb 28, 2013 [cited 2015 July 23]. Available from: http://quest.nasa.gov/space/teachers/suited/3outer.html

The Dangers of Spaceflight

Earth is a protective bubble providing a safe and protective environment for human survival. As soon as we step outside of that bubble, a myriad of challenges and dangers exist. Most of them are due to the harsh environment of space, but a few are due to our own technical shortcomings or are the result of a lack of foresight over the past decades, such as the accumulation of space debris in Earth orbit.

Risks for future spaceflight, including the uncertainties of new technologies and propulsion systems, navigation and guidance accuracies, and an inhospitable space environment will affect both the spacecraft and any humans aboard. The general public understands that risk is necessary to achieve scientific goals and to explore the unknown, much like the ancient explorers who discovered a new world.

Generations have been waiting for the next bold step beyond stepping foot on the Moon. Interest grows and wanes periodically (dictated in part by funding and politics), with always a core group of people worldwide who advocate for space endeavors, accepting the risks as challenges. The purpose of this chapter is to detail the major environmental risks encountered when exploring outer space. Technological challenges will be discussed in a later chapter.

In the early manned space programs of the United States and the USSR, very little was known about the detrimental effects of the space environment on both the spacecraft and human crews. Some of the greatest unknowns were whether or not a human could survive the acceleration required to be launched into space and back, the effects of microgravity on the human body, and whether the heat shield technology could protect the body from the extreme heat of re-entry.

Microgravity Issues

Gravitational forces ("g's" or "g loads") on an animal or human launched into space were considered to be similar to the forces experiences by the accelerations experienced by a jet pilot performing acrobatic maneuvers. A "g" load or force is the effect of Earth's gravitational acceleration on the mass of an object on the surface of Earth and directed inward toward Earth's center. A single "g" load results in the weight of our body, which keeps us attached to Earth's surface. Multiple "g" forces result in multiples of our weight, which pushes down on the entire body, resulting in possible blackout situations caused by the heart not pumping enough blood to the head. Acceleration of a vehicle into space and deceleration returning to Earth's atmosphere results in "g" loads several times that which occurs from the normal acceleration due to gravity. An animal or a human would then weigh several times its normal weight on Earth, putting a strain on body organs.

After entering orbit, microgravity would be experienced, which feels essentially the same as no gravity, releasing all of the pressure that previously was pushing down on the body. After some period of time in this state, the spacecraft would begin deceleration through Earth's atmosphere in order to de-orbit. The body would then experience high deceleration forces after spending a period of time (hours or days) in weightlessness.

Research on "g" forces began decades ago during the early stages of aviation medicine. Initial Mercury testing was done aboard a spacecraft capable of parabolic maneuvers that would result in short periods of time of zero gravity followed by increased acceleration. Other testing was accomplished on the ground using a rocket-powered impact sled, which showed that animals and humans could experience extremely high g's for a fraction of a second. Human tolerance using this device was estimated to be 83 g's for deceleration (experienced for only 0.04 of a second).[4]

Another important testing device was a centrifuge, which looked at the limits of human endurance for both acceleration and deceleration. The centrifuge was a large mechanical arm with a seat at the end of the rotating arm to carry a human or animal. The arm would be rotated at high angular velocities, gradually building up "g" forces. Reactions to the increased forces were chest pain, shortness of breath, and even blackout. Human subjects could tolerate a level of 3 g's for almost 10 minutes and 10 g's for 2 minutes. Prolonged exposure to high g's were found to be damaging to the human body or even fatal. It was determined that 8 g's represented an acceptable safety limit for a human in space.[5] Alan Shepard experienced a maximum acceleration of 6.5 g's during launch and endured a maximum deceleration of 11.6 g's coming back into the atmosphere. His training had including up to 12 g's prior to flight.[6]

G-force testing continued using new methods and devices. Results from spaceflights contributed to spacecraft design, positioning, and design of astronaut seating, and mission design of an optimum re-entry angle. Designs were geared towards a maximum of 9 g's.[7]

---

[4] HQ.NASA. [Internet]. history.nasa.gov. Multiple G; c2015 [cited 2015 Aug 14]. Available from: http://history.nasa.gov/SP-4201/ch2-4.htm

[5] HQ.NASA. [Internet]. history.nasa.gov. Multiple G; c2015 [cited 2015 Aug 14]. Available from: http://history.nasa.gov/SP-4201/ch2-4.htm

[6] White, J. Terry. The flight of Freedom 7. Seattle PI. 2010 May 03.

[7] HQ.NASA. [Internet]. history.nasa.gov. Multiple G; c2015 [cited 2015 Aug 14]. Available from: http://history.nasa.gov/SP-4201/ch2-4.htm

As space missions became longer, human physiology under various gravity conditions were studied more extensively. In addition, new technologies limited severe accelerations. The Space Shuttle design relied on controllers to keep the acceleration of the vehicle lower, no greater than three "g's" to minimize stress on astronauts, eliminating the high "g" force issues. Now, the main issue became the human physiology in microgravity for extended periods of time.

The US and Russian space programs have decades of data regarding the effects of an almost zero gravity environment on the human body. It has long been known that if humans spend an extensive amount of time without gravity, there will be significant effects on human organs, muscles, blood flow and bones. Living and working in microgravity is mentally as well as physically challenging. Gravity on Earth works our muscles to support our bodies, to move, and do physical work. Without gravity, astronauts can experience muscle atrophy, bone degradation, and acceleration of some age-related body changes. However, it was found that many of these effects are essentially reversible a short time after returning to Earth. If humans live in a low gravity environment while they establish a colony, the long-term effects on them are difficult to predict. And, if a human lives out his life on another planet that has low gravity, we will need to know how other planetary conditions affect reproduction, growth, disease, and death.

NASA and other space companies require more studies on long term effects of the issues described here to give them more confidence going forward with deep space mission planning. In March 2015, NASA astronaut Scott Kelly arrived at the International Space Station (ISS) to begin a year-long mission and provide an opportunity for scientists to track long-term bodily changes resulting from a microgravity environment. Scott has a twin brother, Mark, a four-time shuttle astronaut, and someone whose physiology can be compared directly to Scott's.

"Months in space are known to cause loss of bone and muscle mass, weakened immune systems and impaired vision, but the upcoming mission will track those health impacts over time, with better medical technology than Russia's previous year-plus missions," *USA Today* reported.[8]

Changes to the human body due to microgravity concern scientists trying to determine how to keep humans healthy in space and upon return to earth. Let's examine a few ways that the body is affected by a lack of gravity over time.

---

[8] Kelsey D. New mission to space offers special opportunity to track astronaut health. Deseret News. 2015 Mar 28.

Muscle Atrophy

Muscle atrophy is a deterioration of the muscle mass from lack of gravity. It is similar to muscle loss from disuse when someone is disabled or incapacitated. The loss of muscle mass affects strength and endurance, more often than not, in the lower body. These changes in muscle performance can affect the astronauts' ability to perform demanding activities, putting them at a greater risk of injury and fatigue. Even on short spaceflights, lasting 5–11 days, studies show that astronauts can experience up to a 20 % loss of muscle mass.[9]

In past missions, these negative effects were counteracted by spending an extensive time exercising with special equipment. Astronauts on the ISS spend two and a half hours a day exercising to keep their muscles active and strong. Even though muscle mass and strength can be regained once astronauts have returned to Earth, maintaining muscle in space is a concern, especially for long-duration space missions. Currently, the only known way to alleviate muscle atrophy in space is through intensive resistance training, aerobic exercise, along with a good diet.[10]

Bone Loss

Humans on Earth normally experience a loss of bone mass at a rate of 2–5 % a year.[11] Studies show that some women, after menopause, experience bone loss at a rate of 1–2 % higher than normal.[12] In space, however, the loss caused by living in microgravity occurs at a rate of 1–2 % a month. To make the matter worse, the rate of bone production also decreases in space. Some areas, such as the pelvic region, experience more severe bone loss and decreased strength. These issues are linked to increased bone fractures. Studies also suggest that the healing process is diminished under microgravity conditions.[13] After returning to Earth, much of the bone loss is reversible; however, it depends on the amount of time spent in microgravity. So how does this apply to colonization or does it?

---

[9] NASA Information, Lyndon B. Johnson Space Center. Muscle atrophy.

[10] Sutton, Jeffrey. Mar 2015. Celestial Influence. Scientific American. 312:3.

[11] NASA, [Internet]. Science.nasa.gov; c2011. Bones; Oct 01, 2001 [cited 2016 Feb 02]. Available from: http://science.nasa.gov/science-news/science-at-nasa/2001/ast01oct_1/

[12] NIH: [Internet]. NIH.gov. Osteoporosis and related bone diseases national resource center. Aug 2014. [cited 2016 Feb 12]. Available from: http://www.niams.nih.gov/Health_Info/Bone/Osteoporosis/osteoporosis_hoh.asp

[13] Sutton, Jeffrey. Aug 15 2005. How does spending prolonged time in microgravity affect the bodies of astronauts? Scientific American. 290:1. pp. 109–109.

Researchers are working on small devices that simulate gravity. This will hopefully tell us how cells and tissues respond in microgravity. In addition, there is a focus on developing drugs that can decrease the bone loss and/or build up bones.[14]

Studies have also shown that bone loss caused by microgravity can lead to an increased risk for kidney stones, due to an increased amount of calcium emission. There may be other issues contributing to the kidney stone formation, such as decreased urine output and urine acidity. Potassium citrate supplements have been shown to help prevent kidney stone development.[15]

Due to a lack of stress from gravity or muscles, the spine can become misaligned, and discs can swell, having taken on increased water. These changes can cause injuries to the discs after the person returns to Earth. Under normal gravity conditions, when standing, the spine is vertical, and gravity compresses the discs, squeezing out the water. When sleeping, the gravity load is removed, and the discs rehydrate and take on water again. This cycle is important to maintain a healthy spine. Several astronauts have reported increased back pain both during and after a flight. During a flight, their spinal discs continue to swell, resulting in an increase in height because there is no counteracting gravity part of the cycle. The spine becomes stiff, increasing the back pain. Disk herniation is possible upon a return to Earth. Spinal issues are being studied by Jeffrey Lotz, Ph.D., professor and director of the Bioengineering Lab at the University of California. He is funded by NASA to study spinal changes in twelve crew members who have spent 6 months on the International Space Station. He hopes to finish his study in 2018 and apply the results to a Mars mission.[16]

"The take-home message is that microgravity is a fairly significant exposure that changes the biomechanics and biology of the spine. We will be learning a lot about how these changes may relate to back pain symptoms and disc herniation risk. Hopefully this study will lead to countermeasures that prevent back problems from long-duration spaceflight and also help develop treatments for people on Earth," Lotz said.[17]

---

[14] Faulk K. Crew's health a concern on long mars trip/muscle and bone loss, cataracts during 30 months away from earth studied by NASA. Houston Chronicle. 2007 Jan 14; Sect. 3.

[15] NASA: International Space Station—expedition; three science operations status report for the week ending Aug. 15, 2001. M2 Presswire. 2001 Aug 15; Sect. 1.

[16] Sayson, et al. May 2013. Back pain in space and post-flight spine injury: Mechanisms and countermeasure development. Acta Astronautica. 86: pp. 24–38.

[17] Healio. Spine Surgery Today. [Internet]. Healio.gov; c2014. Microgravity negatively affects vertebral disks of astronauts during space flight; May/June 2014 [cited 2015 July 16]. Available from: http://www.healio.com/spine-surgery/disc-biology/news/print/spine-surgery-today/%7Be2cd0249-e4df-4e78-8ec2-e743bf77ea3e%7D/microgravity-negatively-affects-vertebral-discs-of-astronauts-during-space-flight

Radiation

Human beings are all exposed to some level of radiation every day—being close to power lines, using our cell phones, even sitting in the Sun. But in our world, the low dosage effects are repaired by the cells in our body. We can also look at the extreme effects on humans by the bombs dropped on Hiroshima and Nagasaki, Japan, to end World War II. Radiation in space falls somewhere in the middle of these two examples and affects not only humans but the spacecraft itself, often interfering with communications and interacting with electronic circuitry.

Radiation is excess energy that comes from the breakdown of unstable atoms. This energy travels at very high speed, most at the speed of light, in the form of particles or electromagnetic waves or photons.

There are two types of radiation: ionizing (low energy) and non-ionizing (high energy). Non-ionizing radiation includes harmless waves such as visible light, heat, radio waves, microwaves, or radar, which exist all around us. This type of radiation passes through matter without disturbing or breaking bonds or removing electrons from atoms. Ionizing radiation has enough energy to remove electrons from atoms, resulting in charged particles. This process can create a highly unstable atom if electrons are removed from the innermost orbit of the atom. Examples of ionizing radiation are X-rays and cosmic rays (particles that collide with Earth from anywhere outside of the atmosphere). Radiation in space is the ionizing radiation type, which includes the solar wind and its solar particles, gamma rays, X-rays, cosmic radiation (high energy sub-particles possibly originating from supernovae), and trapped radiation in the Van Allen Belts.[18]

The effect of radiation on humans is a complex issue and, therefore, remains an ongoing area of research. What happens to human cells depends on several things—the intensity of the exposure, the particle energy, the time of exposure, and possibly even the person's current physical condition and gender. Extended exposure to ionizing radiation can damage cells, altering its DNA molecules or other important components of cells in control of replication or other cell functions. Cells can die or mutate because of this damage and affect tissues and organs throughout the body, resulting in illness and, in some cases, cancer. Communication between cells can also be altered, promoting behavior that doesn't normally occur, such as accelerated growth of cancerous tumors.

[18]Teodorescu, H. & Globus, A. 2005. Radiation passive shield analysis and design for space applications. SAE Technical Paper 114:179–188.

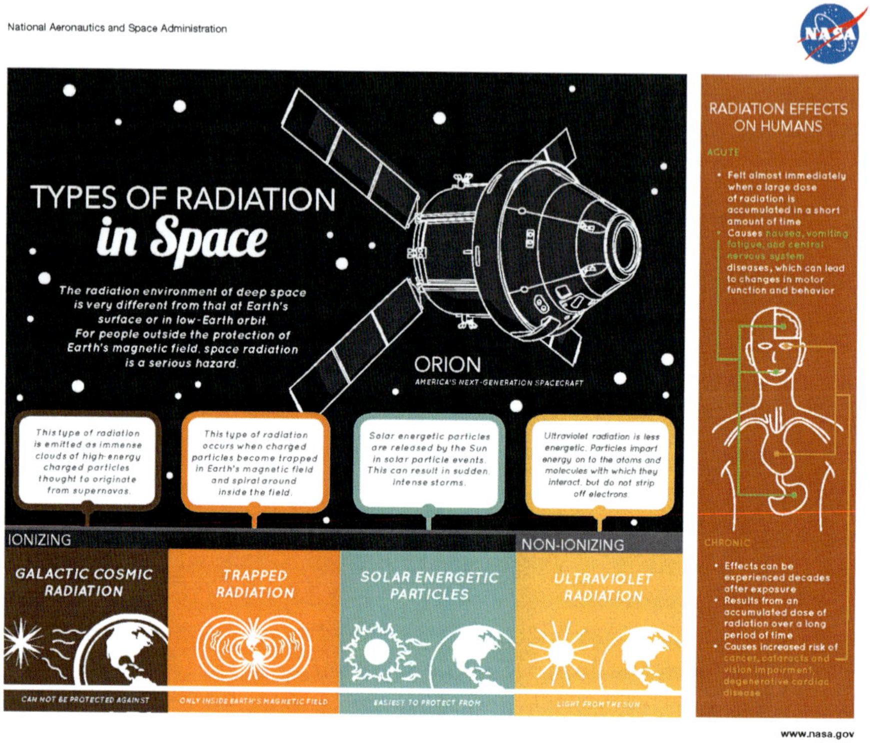

Fig. 5.1 Infographic on radiation in space (Image courtesy of NASA)

Radiation symptoms can range from nausea to vomiting to hair loss to fatigue to death depending on the time between exposures and the severity of the dosage. In addition, ionizing radiation exposure can affect electronic equipment by causing a build-up of static charge over time which could be released and cause damage or failure (Fig. 5.1).

For the most part, Earth is shielded from harmful radiation in outer space by a mysterious force field, much like the fictional *Star Trek* shield that worked to block alien weapons from hitting the starship. Molten iron at Earth's core that spins with the Earth generates a strong enough magnetic field to deflect the flow of particles from us for nearly all of the dangerous forms of radiation. The charges are trapped in two large areas surrounding Earth called the Van Allen Belts. The belts were discovered in 1958, but much of their behavior is still mysterious. During a solar event, the surrounding area outside of the Van Allen Belts becomes saturated with high energy charged particles. Solar events, such as a solar storm resulting from solar flares and coronal mass ejections, causes a solar wind of charged particles that can bombard Earth's magnetic

Fig. 5.2 The Van Allen Belts surround Earth (Image courtesy of NASA GSFC)

field. It just produces a beautiful light show, or stronger events can interrupt electrical grids and damage satellites, even causing satellite phones to go quiet. In 2012, observations from the Van Allen probes showed that a third belt can sometimes appear. In Fig. 5.2 the radiation is shown in yellow, with green representing the spaces between the belts.[19]

Traveling through the Van Allen Belt into outer space may be a problem due to the high concentration of charged particles in some regions. The altitudes of the belts vary—the center of the inner belt is approximately 1860 miles (3000 km) above Earth's surface, and the center of the outer belt is approximately 9300–12,400 miles (15,000–20,000 km) above Earth, although some estimates extend out to 23,700 miles (38,000 km).

Most of the low Earth orbit activities are well outside of this range. The ISS is stationed at approximately 240 miles (390 km) from Earth, well below the lower altitude, and the space shuttle missions varied between 155 miles (249 km) to as high as 600 miles (965 km) from Earth based on mission requirements. Geosynchronous (approximately 22,236 miles, or 35,786 km) communications satellites orbit just inside the outer edge of this radiation belt.[20]

---

[19] NASA Science. Science News. [Internet]. Science.nasa.gov; c2014. Van Allen probes discover new radiation belt; Feb 28, 2013 [cited 2015 July 09]. Available from: http://science.nasa.gov/science-news/science-at-nasa/2013/28feb_thirdbelt/

[20] Editors of Encyclopedia Britannica Online. [Internet]. Britannica.com; c2014. Van Allen radiation belt; [cited 2015 July 09]. Available from: http://www.britannica.com/science/Van-Allen-radiation-belt

Professor James Van Allen and Space Travel

Professor James Van Allen, a brilliant physicist interested in cosmic rays (high energy particles originating outside of the Solar System) known to exist since the 1930s, first started studying cosmic particles from within the atmosphere using sounding rockets and small rockets launched from balloons. In 1955 he was asked to design instruments to be carried on America's first Earth satellites, during the International Geophysical Year (1957–1958). This timeline coincided with peak solar activity. Van Allen initiated cosmic ray and radiation investigations as the first research goals for Earth satellites. Little was known about cosmic rays; however, the increasing levels of radiation with altitude suggested serious hazards for humans in space.

Professor Van Allen put instrumentation onboard early generation satellites (such as Explorer 1 and Pioneer 3) to measure energized particles before they enter the atmosphere. After many satellite studies and extensive analysis, Van Allen verified that clouds of a particular shape composed of high energy particles existed near Earth. The clouds were thickest at the equator and thinner at the poles. The Explorer satellites equipped with shielded instruments (Geiger counters) mapped out two belts of high radiation. These belts that circle Earth were named after Van Allen. They exist in the beautiful auroras in the sky. More importantly, the belts were an important piece of the puzzle in solving how to deal with the hazards of human travel in space. Van Allen then played a major role in designing scientific satellites for NASA. He was awarded the NASA medal of achievement in 1974 plus other prestigious awards, including the Governor's Science Medal, National Medal of Science, and the Gold Medal of the Royal Astronomical Society, for his scientific discoveries. He died at age 91 in 2006.[21]

In 2012, a third radiation belt was measured with probes launched by NASA. These data are leading to a new understanding of the Van Allen Belts and their behavior over time.[22]

Radiation and Space Missions

The first missions to travel outside of low Earth orbit and fly near or through the Van Allen belts were the Apollo missions to the Moon. After their discovery, it was thought that travel through the belts would be problematic

[21] Tucker A. Obituary: James Van Allen: Pioneering physicist who discovered the asteroid belts that bear his name and played a key role in US space exploration. The Guardian. 2006 Aug 11; Sect. 38.

[22] NASA Science. Science News. [Internet]. Science.nasa.gov; c2014. Van Allen probes discover new radiation belt; Feb 28, 2013 [cited 2015 July 09]. Available from: http://science.nasa.gov/science-news/science-at-nasa/2013/28feb_thirdbelt/

and dangerous. NASA needed more information about the radiation belts and their locations to be able to determine the best trajectory for a spacecraft to take to and from the Moon. Van Allen worked with Werner Von Braun (director of NASA's Marshall Space Flight Center and architect of the Saturn V launch vehicle) on the hazard of extended manned missions, concentrating specifically on Apollo. It was determined that Apollo could choose orbits that would avoid the strongest levels of radiation and that would allow the spacecraft to quickly travel through the radiation from Earth to the Moon, eliminating any long-term effects. These preventative measures kept the radiation doses low.

Additional studies were done on the consequences of a solar flare (a brief and intense eruption of high energy radiation from the surface of the Sun) during missions that would increase the charged particle flux. The astronauts were issued handheld Geiger counters to assist them in determining the best place in the command module for shielding in case of a solar flare. No significant solar events occurred during the Apollo missions.[23]

Radiation for long-term missions such as the mission to Mars represents a bigger challenge for human exposure. Data collected by the Curiosity rover mission measured daily excessive radiation exposure equivalent to a 5 % increase in fatal cancer risk. This didn't include a long-term stay on the planet.[24] Astronauts on a Mars mission would exceed their lifetime radiation exposure limits at 18 months to 2 years travel in space. This increases the radiation threat and the urgency for a solution within a decade.[25]

Some variations in the solar cycles can be predicted by scientists helping in the planning of a mission timeline. In addition, some types of medicines could help prevent cell damage from cosmic rays. However, the unknowns in terms of how much radiation is too much are still being studied. In space, the high energy particles are everywhere and moving so fast that it is difficult to stop them by shielding, even though this is an obvious solution to prevent particles from reaching the human body. Wave radiation traveling at high speeds in outer space requires thick shielding. Thick could mean dense and heavy, making such materials difficult or too expensive to transport or create a spacecraft out of. In order to shield space vehicles and other structures, new

[23] Earl Lane. Newsday. Scientists studying space radiation, ways to fight effects. Las Vegas Review—Journal. 1991 Dec 29; Sect. 10c.

[24] Allen, Kate. [Internet]. thestar.com; c2013. Another hitch on the way to Mars: too much radiation; May 30, 2013 [cited 2015 July 11]. Available from: http://www.thestar.com/news/the_world_daily/2013/05/another-hitch-on-the-way-to-mars-too-much-radiation.html

[25] NASA Fact Sheet. Lyndon B. Johnson Space Center. Understanding space radiation. FS-2002-10-080-JSC. Oct 2002.

technologies or materials have to be developed in order to protect against the ionizing effect of space radiation.

NASA awarded prize money in April of 2015 to five physicists to develop innovative ways of protecting crews from dangerous radiation exposure on the journey to Mars. The mission to Mars requires crews to be exposed for 500 to more than 1000 days. Shielding human life has become a NASA priority on long duration missions and exposure to galactic cosmic rays. The goal of this award is to produce methods to reduce crew members total radiation dosage by at least a factor of four. There will be a second challenge asking for ideas to provide maximum protection to shield the crew. Ideas include using magnetic or electrostatic fields to deflect harmful radiation or using layered material to shield the crew.[26]

In December 2014, NASA completed a test flight of the Orion Multi-Purpose Crew Vehicle, a deep-space exploration capsule designed to carry crews to or beyond low-Earth orbit eventually atop the Space Launch System. Twice during the test flight, Orion traveled through the Van Allen Belt, measuring the effect of deep-space radiation on its onboard electronics, which are designed to be radiation tolerant. The military has extensive experience designing equipment to be radiation hardened in case of proximity to bomb detonations. This experience has proved invaluable to space travel, where there is extensive exposure to highly charged particles traveling at high speeds.[27]

Additional discoveries are on the horizon with regard to materials that could block harmful radiation. Measurements taken onboard NASA's Lunar Reconnaissance Orbiter (which has been circling the Moon since 2009) have shown that plastic shielding reduces the radiation dosage from galactic cosmic rays. These discoveries would confirm previous ideas that lightweight materials such as plastic can be as or more effective for blocking cosmic radiation than other materials such as aluminum.[28]

Long duration missions in space will only be possible with extensive developments in radiation shielding. Passive shielding, currently the only method available, is the use of materials that physically block the energetic particles, for example, by using sheets of aluminum. A greater thickness yields more

[26] NASA awards radiation challenge winners, launches next round to seek ideas for protecting humans on the journey to mars. PR Newswire. 2015 Apr 16.

[27] Howard, Courtney. Averting on-orbit million failure. Military & Aerospace Electronics. May 2015. 26:5:20–30.

[28] Wall, Mike. [Internet]. Space.com; c2014. Plastic could protect astronauts from deep-space radiation; June 14, 2013 [cited 2015 July 23]. Available from: http://www.space.com/21561-space-exploration-radiation-protection-plastic.html

protection at the expense of increased mass. Active shielding is a popular idea that has not matured enough for spaceflight.

The first human mission to Mars will require a combination of active and passive methods of shielding and mission planning that focuses on an optimal choice of trajectory for the least amount of radiation exposure. It will require a propulsion system that can deliver faster transit. It could be that a combination of thermal and electric nuclear power might be the answer, but that technology hasn't fully matured.[29]

Cardiovascular Effects

The human circulatory system works against gravity. Gravity pulls blood down to the lower half of the body, and the heart works to pull it back up to the upper half of the body. When there is almost no gravity, the heart doesn't need to work as hard to send blood to the upper body, including the brain. You would think this was a good thing, but in fact it causes the blood to accumulate in the upper body. In this case, the system sends a different set of signals to control and adapt to the new environment. These changes in fluid distribution result in modifications to blood pressure and the amount of blood pumped by the heart with each beat. Since the body doesn't have to work as hard, the heart becomes less efficient and generates slightly highly systolic and diastolic pressures because some of the muscle groups, like the lower legs, are less active and not sending signals for more blood. In addition, inefficiency causes some blood to remain in the heart, leading to a slight increase in the relaxation phase pressure, the diastole. The end result is that the amount of blood pumped out of the heart, referred to as the stroke volume, decreases, and as time goes on, this causes the changes to become more pronounced and possibly affect other bodily functions. It could even result in permanent changes to the way blood vessels and some organs behave. Similar to counteracting muscle atrophy, the astronauts can engage in activities after returning to Earth in order to de-condition the cardiovascular system back to normal.[30]

[29] Durante, Marco. Space radiation protection: Destination Mars. Life Sciences in Space Research. Jan 2014. 1:2–9.

[30] NASA Educational & Texas Instruments. Microgravity effects on human physiology: circulatory system. 2011.

Changes to the Human Mind in Space

It has long been thought that there could be some overwhelming "madness" that might occur if a human were left to experience long periods of microgravity and isolation on a long-term space mission. This hasn't occurred yet, but new problems could appear in future long-term and long-distance space missions. Living in space is demanding and stressful. Mental stresses, such as worries about family or friends or being alone and distant from Earth, can adversely affect a crew member's health, work performance, and overall welfare. Astronauts are subjected to extensive psychological testing prior to their selection process, and going forward, this process will be beneficial in minimizing the chances of behavioral problems and psychiatric disorders in space.

Thus far, the only documented issues involved mood and personality conflicts among crew members in space were specifically observed on the ISS where both American and Russian crewmembers interacted with each other. Some of the issues were determined to be the result of cultural differences and others resulted from program dissimilarities and the way that daily mission activities are carried out. Yuri Gagarin, cosmonaut and first man in space observed: "Without a doubt, in our country it is much easier to form a crew for a long-duration space mission than in capitalist countries. [We] are collectivists by nature." [31]

It is also thought that the psychological issues identified on the ISS would be magnified for a longer mission such as the mission to Mars. Longer term missions would lack the structured communication to mission control and would provide the additional mental stress in the knowledge of being far away from Earth and contact with family, friends, and colleagues. Watching Earth become smaller and smaller could increase feelings of isolation and anxiety. Some of these reactions are difficult to simulate prior to the mission. The only steps that can be taken are to take as many precautions as possible to ensure the most stable candidates are chosen for long term missions.[32]

Radiation exposure has been associated with a rise in physical ailments, but long term exposure to radiation can also have an effect on mental diseases. Studies show that increased levels of radiation, such as the level of exposure that would be experienced on a mission to Mars, could cause problems with reasoning and accelerate the onset of Alzheimer's disease. Alzheimer's is a

[31] Ritsher, Jennifer Boyd. 2005 Cultural factors and the international space station. *Aviation Space and Environmental Medicine*, 76(6), 135–144.

[32] Kanas, et al. Apr 2007. Psychosocial interactions during ISS missions. Acta Astronautica. 60:pp. 329–335.

neurodegenerative disease that is enhanced by the effects of ionizing radiation. Astronauts can be shielded from radiation associated with solar flares given warning, but other forms of cosmic radiation cannot be blocked at this time due to lack of available materials (use of lead walls result in very heavy spacecraft). Studies are still being conducted to investigate both the biological and cognitive effects of ionizing radiation on the brain.[33]

If a disease like Alzheimer's can be accelerated due to ionizing radiation, it is logical to think that parts of the brain could be damaged due to radiation exposure. Results from studies have indicated that long term exposure to charged particles may result in a decay of brain functioning, which in turn could affect decision making and severely impact the success of a space mission. Recent evidence shows that radiation has the capability to compromise the integrity of neurons throughout the brain, contributing to degenerative functioning. More studies are being conducted to further study the effects on brain functioning.[34]

Space Debris

If you use one of your sky guide apps on your phone and hold it up to the night sky, you can see manmade space junk identified as it moves across the sky. Space debris also includes both natural objects such as meteoroids in orbit about the Sun. Most artificial or manmade debris is in orbit around Earth. These objects are commonly referred to as orbital debris or space junk.

Satellites and rockets have been launched for decades, starting with Sputnik in 1957. Satellites have been propelled into space by the hundreds over the years, and when they become inoperative, they are left as space garbage. The collection of space debris also includes the remains of past missions, such as depleted rocket stages and lost astronaut equipment, such as a stray glove floating around. It is estimated that more than 500,000 pieces of debris can be tracked orbiting Earth.[35] They are all traveling at a minimum speed of 17,500 mph, which is the minimum speed for an orbiting object and certainly fast enough to damage or destroy a passing spacecraft or satellite. At that speed,

[33] Begum, et al. Nov 2012. Does ionizing radiation influence Alzheimer's disease risk? Journal of Radiation Research. 1:53(6):815–822.

[34] Parihar, et al. May 2015. What happens to your brain on the way to Mars? Cognitive Neuroscience. 1:4:e1400256.

[35] NASA, [Internet]. Science.nasa.gov; c2014. Space debris and human spacecraft; Sep 26, 2013 [cited 2015 July 26]. Available from: http://www.nasa.gov/mission_pages/station/news/orbital_debris.html

even a half-inch piece of debris would have the kinetic force of a bowling ball thrown 300 miles per hour, according to NASA.[36]

Significant threats of collisions between satellites or spacecraft have been reported, but actual collisions of spacecraft with space debris have not yet happened to the extent shown in the movie *Gravity*. In that movie, a large amount of debris from a collision destroys a spacecraft, leaving its crewmembers either dead or stranded. However, in real life, there has been property damage and potential threats to human life.

"There are up to 200 threats a day identified for orbiting satellites," said Trevor Thomas, a Lockheed Martin spokesman. "Most satellites can sustain some damage, but little bits of junk hit satellites every day, and each [satellite], on average, is worth around $500 million."[37]

Orbiting in an environment with little atmosphere and only a small amount of gravitational pull, any spacecraft, even those deemed inoperative, can remain in space for a long time. At an altitude of under 500 miles (805 km) there is sufficient atmospheric drag to eventually pull the spacecraft back to Earth, but only after many years, even decades. Until then, unless there is an onboard mechanism for initiating a de-orbit burn, the craft remains as an unusable piece of space junk. Up to 75 % of all debris is located in low Earth orbit, the most popular regions for satellites and manned spacecraft (the region containing altitudes from about 99 miles, or 160 km, to 1200 miles, or 2000 km) above Earth's surface.[38] Anything below that altitude will quickly decay due to Earth's gravity and atmosphere and be pulled back toward Earth.

Figure 5.3 shows the number of debris objects by altitude for the year 2012. Two spikes occur in the peak associated with the majority of satellites located close to an 800 km orbit. These spikes are associated with events that resulted in a large number of small debris particles. One of the events identified was the result of a collision between two communication satellites over northern Siberia on Feb. 10, 2009. The impact of the Iridium 33 (US) satellite and a non-working Soviet Union-era satellite (Cosmos 2251) resulted in hundreds

[36] Zenko, Macah. FP. [Internet]. Foreignpolicy.com; c2014. 135 million pieces of junk are orbiting Earth at 18,000 mph—and U.S. space dominance is in danger of being ripped to shreds.; 21 Apr 2014 [cited 2015 July 27]. Available from: http://foreignpolicy.com/2014/04/21/waste-of-space/

[37] Taylor R. Space junk: New tactics to curb risk. The Wall Street Journal Asia [Hong Kong]. 27 Aug 2014:1.

[38] Zenko, Macah. FP. [Internet]. Foreignpolicy.com; c2014. 135 million pieces of junk are orbiting Earth at 18,000 mph—and U.S. space dominance is in danger of being ripped to shreds.; 21 Apr 2014 [cited 2015 July 27]. Available from: http://foreignpolicy.com/2014/04/21/waste-of-space/

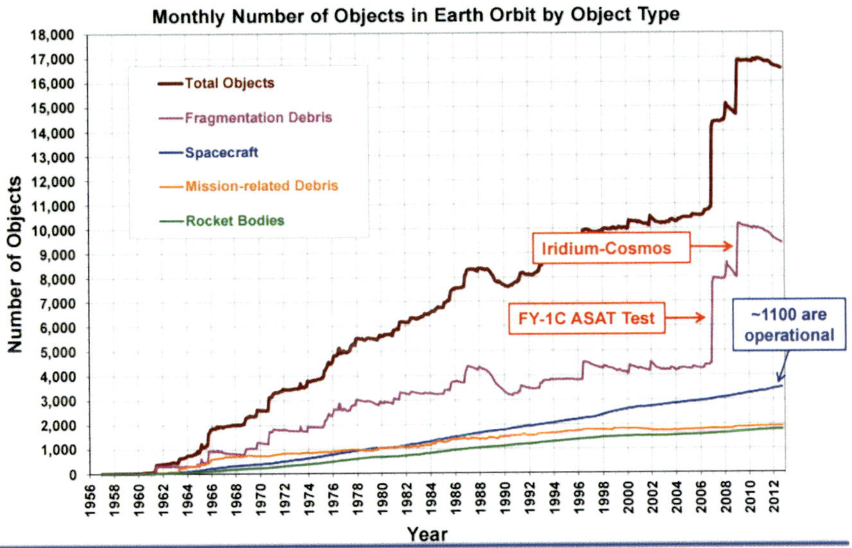

Fig. 5.3 Growth of orbital space objects including debris (Image courtesy of NASA)

of thousands of fragments, most of which will eventually decay back to Earth but remain factors in potential impacts with other spacecraft for decades.[39]

The second event was the breakup of the non-operational Chinese meteorological satellite Fengyun-1C on January 11, 2007, as it flew over central China, which was intentionally destroyed by a ballistic missile kinetic kill vehicle (KKV) launched from the Xichang Space Launch Center. The KKV travels through space at over 32,000 km/hour following a ballistic arc rather than enter into orbit. The destruction created thousands of pieces of debris, again much of it that will remain in orbit for decades. These particles resulted in a ring of debris in a similar orbit as the destroyed satellite, traveling in the same direction and velocity. This became the largest debris cloud ever generated by a single event in orbit (Fig. 5.4).[40]

In late March 2016, the Japan Aerospace Agency (JAXA) lost contact with its $273 million satellite (Hitomi) only a month after it had been launched. The high-tech X-ray observatory was developed in partnership with NASA

[39] David, Leonard. [Internet]. Space.com; c2014. Effects of worst satellite breakups in history still felt today; June 28, 2013 [cited 2016 Feb 15]. Available from: http://www.space.com/19450-space-junk-worst-events-anniversaries.html

[40] Johnson, Nicholas et al. [Internet]. NASA.gov archives. The characteristics and consequences of the break-up of the Fengyun-1C spacecraft. 2007. IAC-07-A6.3.01 [cited 2016 Feb 15]. Available from: http://ntrs.nasa.gov/archive/nasa/casi.ntrs.nasa.gov/20070007324.pdf

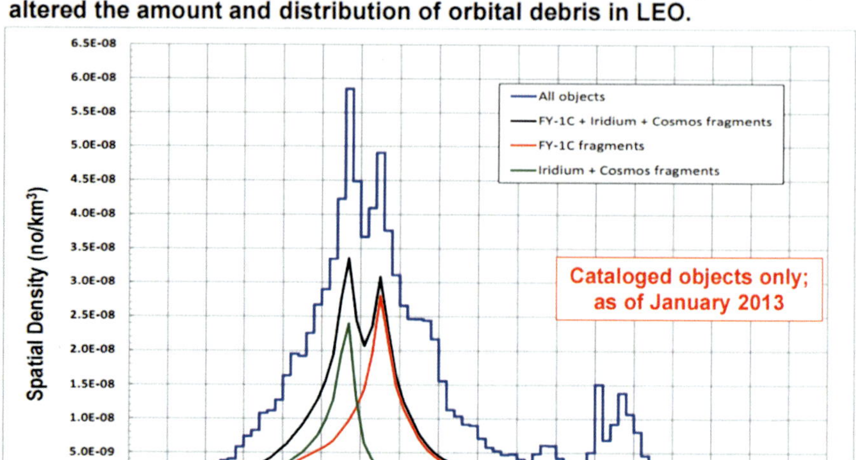

Fig. 5.4 Space debris density by altitude above Earth (Image courtesy of NASA)

to observe and collect data on some of the biggest mysteries of the universe, primarily high energy particles that emanate from black holes, supernovae, and exploding stars. The U.S. Joint Space Operations Center identified and tracked five objects at the satellite's location when it lost contact with Earth, indicating that the craft broke up or was hit.[41]

The rising number of objects in orbit increases the danger to all space vehicles, including the ISS and other human occupied spacecraft, as well as private and government satellites providing vital communication and surveillance services. Disruption of these services becomes a real possibility as the probability of collisions increases. In addition, damage to satellites is generally impossible to repair now that we don't have the space shuttle orbiter available to rendezvous with the wounded craft and repair it. In addition, manned spacecraft are difficult to repair and involve the ability to do spacewalks to patch or fix damaged areas and equipment. The preference is to avoid the threat if possible or take defensive measures to protect the spacecraft. Advanced warning helps to determine a course of action to save a spacecraft or a life.

[41] Calderone, Julia. Japan has lost a recently launched space satellite. Where could it be? The X-ray satellite, Hitomi, is the third in a series of ill-fated space observatories. The Christian Science Monitor. 28 Mar 2016.

The space debris problem can become politically complex when it appears that space collisions happen intentionally, such as China's 2007 missile test that destroyed an old satellite. After this event, Chinese officials said they would not conduct further tests; however, similar tests were conducted in 2010 and 2013 under the description of missile defense. It has been determined that China has made significant developments in their capabilities to disable or destroy satellites, which is of concern to U.S. officials and other countries with space programs that have an interest in investing in spacecraft for low Earth orbit.[42]

Monitoring space debris is a huge and complex task. The U.S. Air Force Space Surveillance, a radar system known as the "Space Fence," was operational from 1961 to 2013, when it needed a technological upgrade. The system could detect space objects, meteors, and debris about the size of a basketball at 30,000 km. Objects were cataloged and used if necessary for collision avoidance. The Joint Space Operations Center (JSpOC) at Vandenberg Air Force Base maintains the US space catalog and combines and analyzes data from other sources to integrate the overall view of Earth orbiting spacecraft. There has been some concern about the capability of tracking objects during the period of time when the Space Fence is not operational. Stopgap measures include radar surveillance systems that seem to be performing adequately in the interim.[43]

Plans for a new Space Fence are underway with a goal of becoming operational by 2017. In 2014 Lockheed Martin Corp. was awarded a nearly billion dollar contract by the US Air Force to develop the new technology space surveillance system. This system will have the capability of tracking as many as 200,000 pieces of orbiting debris circling in low Earth orbit. The adapted technology uses optical and laser tracking first tested on the battlefields of Iraq and Afghanistan to locate debris moving at speeds of 17,500 mph (28,200 km/hour).[44]

"Previously, the Air Force could only track and identify items the size of a basketball," said Dana Whalley, the government's program manager. "With the new system, we'll be able to identify items down to the size of a softball. This will significantly increase our capability."

---

[42] Truong K. Wayward space junk prompts astronauts to shelter in cosmic lifeboat. The Christian Science Monitor. 16 Jul 2015.

[43] Defense Industry Daily Staff. Don't touch their junk: USAF's SSA tracking space debis. Defense Industry Daily. 26 Aug 2014.

[44] Hennigan WJ. Watching over a cosmic minefield; Lockheed's 'space fence' surveillance system will track debris orbiting earth. Los Angeles Times. 05 Jul 2014.

Researchers have categorized more than 23,000 items of the size bigger than a basketball, the tracking system's current resolution capability. In addition, it was estimated that there are thousands of pieces of debris smaller than the size of a baseball that could still damage a functioning spacecraft because of their significant speeds. "The greatest risk to space missions comes from non-trackable debris," said Nicholas Johnson, NASA chief scientist for orbital debris in a statement.[45]

It is estimated that at least nine collisions between non-classified satellites have occurred in the past 50 years. This does not seem like a high number, but the potential risks are growing every year. Every collision results in thousands of small particles that can damage other spacecraft. The number of total objects to track has almost tripled to over 15,000 in 30 years, and with new technology able to detect smaller objects, there will be an increased number of objects to track and catalog. The new Space Fence program will track objects in low to medium Earth orbit and will operate in the shorter wavelength S-band frequency range, replacing the old system using VHF. The architecture will be modernized as well, bringing the detection and analysis to a new level. An estimated 100,000 space objects previously not able to be tracked by the past radar system will be able to be seen with this wider band capability. In addition, better locations around the world for the network sensors will allow a wider region to be scanned.[46]

Warnings for proximity alerts for space debris are issued regularly to the ISS. In July 2015, three astronauts took refuge in an escape capsule (a Soyuz spacecraft) docked to the station in order to protect themselves from a possible collision with a piece of space junk. Luckily, the debris passed by about a mile and a half away. This type of precaution has only been required three other times, and, thus far, no major collisions have been recorded. The ISS is capable of moving out of the way to avoid a collision, but it takes more than a day's time to plan and carry out the evasive maneuvers. As was done in 2015, if an imminent collision is only a few hours away, the best plan of action is to protect the crew inside of escape vehicles.[47]

NASA has developed a set of guidelines used to assess whether the proximity of orbital debris to a spacecraft is sufficient to perform evasive actions or safety measures for a crew. These guidelines draw a "pizza box" shape around

---

[45] Truong K. Wayward space junk prompts astronauts to shelter in cosmic lifeboat. The Christian Science Monitor. 16 Jul 2015.

[46] Defense Industry Daily Staff. Don't touch their junk: USAF's SSA tracking space debis. Defense Industry Daily. 26 Aug 2014.

[47] Truong K. Wayward space junk prompts astronauts to shelter in cosmic lifeboat. The Christian Science Monitor. 16 Jul 2015.

the spacecraft, about 30 miles across by 30 miles long by about a mile deep by (1.5 × 50 × 50 km), with the vehicle in the center. If analysis predicts that debris will pass close enough for concern and the quality of the tracking data is deemed sufficiently accurate, Mission Control centers in Houston and Moscow work together to develop a course of action.

The analysis is based on the calculation of the probability of a collision based on the miss distance and an uncertainty provided by JSpOC. In making this critical decision, JSpOC would collect additional tracking data on the threat to improve accuracy in the prediction. If a maneuver is necessary, NASA provides JSpOC data regarding the new orbit for future predictions.[48]

To deal with minor debris, the ISS is fitted with a thin shield called a Whipple Bumper. The shields were first invented by Fred Whipple in 1946, who called it a "meteor bumper." It works in a similar way as a car bumper in that a thin outer shield, such as thin sheet of aluminum, is separated from the spacecraft by an open space and works to shield the craft and cause the small pieces of debris to explode when they strike the surface. The shields have also been stuffed with layers of material that can absorb shock in a similar way that crash walls are constructed on race tracks. Many types of shields have been designed and are used on all types of spacecraft. The ISS alone used 200 Whipple-type shields to protect from impacts.[49]

Collision avoidance and bumpers are not the only methods being considered as preventative measures for orbiting space debris. Lasers are being explored as a means of slowing objects down so they return back to Earth, burning up on the way. There is even an idea being considered to outfit the ISS with a type of laser cannon developed by Tokyo researchers. Plans for its testing were recently released in May 2015. The high-powered laser, which could be considered a weapon, would eventually be able to push debris back into Earth's atmosphere.[50] Shooting lasers into space has its own risks, and questions might arise as to the nature and purpose of the laser as well as the legality of shooting an object that officially belongs to someone else.

Other proposed methods include a spacecraft that would be deployed as sort of a garbage truck, picking up smaller pieces of debris and non-working satellites. The space shuttle orbiter would have been a good candidate for this

[48] NASA, [Internet]. Science.nasa.gov; c2014. Space debris and human spacecraft; Sep 26, 2013 [cited 2015 Aug 09]. Available from: http://www.nasa.gov/mission_pages/station/news/orbital_debris.html

[49] Innovators. The Ottawa Citizen. 02 Jan 2005; Sect. C14.

[50] Prigg, Mark & O'Callaghan, Jonathan. [Internet]. Dailymail.co.UK; c2015. The real Death Star! International Space Station to get a laser cannon to shoot away orbiting junk.; May 19, 2015 [cited 2015 Aug 07]. Available from: http://www.dailymail.co.uk/sciencetech/article-3088370/The-REAL-Death-Star-International-Space-Station-laser-cannon-shoot-away-orbiting-junk.html

type of mission, but now that the program has been canceled, there is no similar American vehicle at this time. Russia has proposed using a nuclear powered spacecraft for long duration missions to pick up or destroy objects. However, more powerful nuclear propulsion systems have not yet been developed and tested and would be potentially dangerous if something went wrong. The positive attributes would be the ability to change orbits and do a variety of work, pulling rocket stages or satellites to lower orbits and clearing debris.[51]

Japan's Aerospace Agency's (JAXA) newest idea is to use a magnetic net to attract and catch metal junk. Japan has joined with a fishing equipment company to create a unique net that will be used to catch some of the orbiting space junk, which is primarily metal. The electrified net, made of ultra-thin stainless steel and aluminum, would first catch and then slow down the debris so it will burn up in Earth's atmosphere. A test launch in early 2015 sent a satellite into space that unraveled the 980 foot (300 m) net into space. The net will be operational in Earth's orbit for about a year, when it will be pulled back to the surface by gravity, incinerating the space garbage on the descent.[52]

Each of these methods has both positive and negative aspects, but it is at least encouraging that a significant amount of effort is going into the development of space junk removal methods. In addition to the removal of space debris, limiting the amount of orbiting debris and making each country or company responsible for its own spacecraft in orbit is equally important. There is the issue of enforcement and making sure countries have agreed to the necessary guidelines as outlined in the Space Treaty of 1967, where it states "States shall be responsible for their national activities in outer space, whether carried on by governmental or non-governmental entities." In addition, "States shall avoid the harmful contamination of outer space."[53] However, removing space debris today would have to be agreed on by interested parties internationally, and there is some doubt as to whether countries are willing to cooperate in this effort and to what extent. There are several countries interested in protecting their space interests and addressing space debris issues. China has recently established an agency to track and deal with space debris due to the increasing threat to their orbiting space assets.[54]

---

[51] Anonymous. 'Space towboats' to have nuclear engines. Interfax: Russia & CIS general newswire [Moscow]. 11 Feb 2010.

[52] McCurry J. In space, no one can hear you clean: The Guardian. 2014 Feb 28; Sect. 25.

[53] Zenko, Micah. [Internet]. Foreignpolicycom; c2015. Waste of space.; April 21, 2014 [cited 2015 Aug 08]. Available from: http://foreignpolicy.com/2014/04/21/waste-of-space/

[54] Zhao L. Agency set to track, deal with space junk. China Daily (Hong Kong ed.). 2015 Jun 10; Sect. 4.

In 2008, after the Iridium-Russian satellite collision, UN member countries adopted a resolution outlining space debris guidelines that call for the removal of non-working spacecraft from low earth orbit.[55] The success of the agreement depends on which nations support the proposal as well as the means to enforce the agreement. "The prompt implementation of appropriate space debris mitigation measures is in humanity's common interest, particularly if we are to preserve the outer space environment for future generations," says Mazlan Othman, director of the U.N. Office for Outer Space Affairs (UNOOSA).[56] The first step is an international consensus to formulate guidelines to create the sustainable use of outer space and prevent pollution of the space environment. Reaching an agreement on these guidelines makes a statement about how important the issue of space debris is to the international community, not only scientists. The UN guidelines outline mitigation measures that involve all phases of spacecraft, including planning, design, manufacture, and operations. Of importance is limiting the longevity of spacecraft remaining in low Earth orbit well past their mission has ended and removing them from this congested region.[57] Again, to date, these are voluntary guidelines with no enforcement attached to them.

Some scientists feel that we are approaching the Kessler syndrome, a theory proposed by a NASA scientist of the same name in 1978. Essentially the theory states that two objects that randomly collide in space generate more debris that collides with other objects, creating more projectiles causing more random collisions until low Earth orbit is so full of debris that passage through it becomes impossible.[58]

Final Thoughts on the Dangers of Outer Space

In the early manned space programs of the United States and the USSR, very little was known about the detrimental effects of the space environment on both spacecraft and human crews. In particular, one of the greatest unknowns was whether or not a human could survive extreme accelerations and the

---

[55] Robin McKie and MD. National: Warning of catastrophe from mass of 'space junk': Failure to act would be folly, says report to UN. The Observer. 2008 Feb 24; Sect. 25.

[56] United Nations, [Internet]. un.org; c2015. Space debris: orbiting debris threatens sustainable use of outer space.; 2008 [cited 2015 Aug 09]. Available from: http://www.un.org/en/events/tenstories/08/spacedebris.shtml

[57] United Nations, [Internet]. un.org; c2015. Space debris: orbiting debris threatens sustainable use of outer space; 2008 [cited 2015 Aug 09]. Available from: http://www.un.org/en/events/tenstories/08/spacedebris.shtml

[58] Sommer M. UB researcher studying space junk. Buffalo News. 2014 Jan 19.

re-entry conditions. Once it was determined that humans could survive space-flight and explore and work in space, attention focused on long-term issues. Going to Mars can take 3 months one way. Additionally, Mars gravity is only 38 % of Earth's surface gravity. A trip to Mars will involve extended periods of time exposed to almost zero gravity conditions. One question that we might ask is whether someone born in a microgravity environment could live in a gravity environment? Specifically, what happens if an astronaut conceives and gives birth on a Mars mission? Would that baby have to remain in the micro-gravity environment or risk debilitation or worse by returning to Earth? These questions are important to address prior to future human colonization or any long-term exploration missions.

In addition to environmental issues, we have seen that orbital debris and asteroid fragments in outer space can destroy spacecraft while at the same time create more debris. Future challenges include addressing the removal of orbital debris and making spacecraft safe from collisions.

Human space exploration requires a commitment to address and solve the important issues that put human beings at risk on long-term missions. The hazards of manned space flight provide new challenges that will be met with future scientific and technological achievements. Also, a commitment to a peaceful and sustainable space environment can only be achieved with the cooperation of international and commercial partnerships.

6

Politics and the Space Race

Keywords Apollo • Cold War • Communism • Cuban Missile Crisis • Douglas Aircraft Company's Project RAND (Research and Development) • Eisenhower • Father of modern rocketry • Yuri Gagarin • Gemini • John Glenn • Robert Goddard • Helmut Gröttrup • Sergei Korolev • Hermann Oberth • International Geophysical Year (IGY) • John F. Kennedy • Konstantin Tsiolkovsky • Laika • Liquid-fueled rocket • Luna • Lyndon Johnson • Mercury • National Aeronautics and Space Administration (NASA) • National Advisory Committee for Aeronautics (NACA) • Nikita Khrushchev • Peenemünde • Richard Nixon • Saturn V rocket • Alan Shepard • Space Race • Sputnik • V-2 rocket • Vanguard rocket • Vostok spacecraft • Wernher von Braun

> "We choose to go to the moon in this decade and do the other things, not because they are easy, but because they are hard, because that goal will serve to organize and measure the best of our energies and skills, because that challenge is one that we are willing to accept, one we are unwilling to postpone, and one which we intend to win, and the others, too."
>
> –President Kennedy (Rice University, 1961)

Scientific discoveries often result from research funded by government organizations, universities, private enterprise, or non-profit organizations. In the request for funding, scientists and engineers have to defend the purpose and application of their proposed efforts or how it will further research in a particular field. Topics of national interest representing economic, military or cultural goals are more often funded than research in other disciplines. Enduring topics that are consistently funded include the health, security, and quality of life of Americans.

L. Dawson, *The Politics and Perils of Space Exploration*,
Springer Praxis Books, DOI 10.1007/978-3-319-38813-7_6

What is troubling to many is that the determination of the most important topics of national interest are decided through the political channels involving proposals pitched by supporters, lobbying style. Some organizations without sufficiently strong supporters run the risk of losing funding to other more aggressive and even celebrity representation. The current budget of 2016 is funding NASA and space-related scientific endeavors at a higher rate in general than was requested. A number of astronauts past and present were quite vocal, appearing before Congress presenting strong cases for the development of transportation spacecraft to the International Space Station while depicting a bleak reality of delaying missions to Mars by years if funding is cut for critical development projects. The result was positive for space science at least in the short term. However, it is disconcerting that funding for the future of space exploration is so dependent on political support and party affiliation and its focus. The main thread throughout this chapter is how funding, politics, and scientific advancement remain closely connected and can negatively affect the future of manned space exploration.

Many Americans believe that space exploration is a challenging and a fascinating pursuit worthy of American government resources. However, there are just as many who believe the opposite—that funding should be expended on other national priorities such as terrorism, the military, environmental issues, and clean energy. Space supporters believe that outer space pursuits distinguish the United States as a global leader and serve to further scientific knowledge critical to understanding the universe and the developing of innovative technologies to be used on Earth for the betterment of our citizens.

There are certain scientific topics that are consistently at the forefront in a US national debate on funding priorities. Climate change and space exploration are two such topics. Scientists continually have to present evidence to support funding for initiatives that not everyone favors or easily understands. Often, the emphasis in national funding is tied to political trends and to the party in control. In order to make funding more stable, consistent public support would help define resources going forward. Grass roots efforts to emphasize scientific discoveries and the benefits of exploring outer space would be one way to get legislators' attention. Recently, a number of expert scientists have been very publicly debating the merits of NASA and the funding of space exploration, both manned and unmanned programs. There also seems to be a renewal of space awareness and enthusiasm due to a variety of new privately funded space ventures and a resurgence of science fiction books and movies.

This chapter examines the national and international politics of space exploration in the early years of space exploration often referred to as the space race.

An Introduction to Modern Rocketry

The French novelist Jules Verne inspired generations with his creative writing and forward-thinking visions of outer space. His 1865 novel *From the Earth to the Moon* foretold the lunar expedition that became a reality a century later. The story was more than science fiction. In fact most of the dialog is scientific and technical, and many aspects of the Moon mission became reality in the Apollo program.[1] Many scientists worldwide were looking to the stars and developing ways of achieving the dream of leaving Earth and traveling into space.

The focus early was rocket development. The key pioneering rocket scientists were the Russian Konstantin Tsiolkovsky (1857–1935), the German Hermann Oberth (1894–1989), and the American Robert Goddard (1882–1945). All contributed to both the theoretical and practical development of rockets. Tsiolkovsky published the rocket equation, defining the relationship between rocket speed and mass and later, a theory of multistage rockets. Hermann Oberth was instrumental in his work on multistage rockets and by contributions to the V-2 rocket for Nazi Germany, bridging the gap to the military uses of rockets. Robert Goddard's focus was on reaching beyond Earth's atmosphere with his development and testing of a liquid-fueled rocket. He also proposed using multistage rockets with solid fuel in addition to liquid fuels. He became the father of modern rocketry in America, but never saw his vision become a reality.[2] His story is told here.

Robert Goddard, the Father of American Rocketry

Dr. Robert Goddard is considered to be the father of modern rocket propulsion. He was a brilliant engineer, physicist, and inventor, growing up dreaming of launching rockets into space. As the story goes, he was inspired by the H. G. Wells' science fiction classic *War of the Worlds*, a work that fueled his fascination

[1] Gioia, Ted. [Internet]. Conceptualfiction.com; c2015. From the Earth to the Moon by Jules Verne; [cited 2015 Aug 28]. Available from: http://www.conceptualfiction.com/from_earth_to_moon.html

[2] Howell, Elizabeth. [Internet]. Space.com; c2014. Rockets: a history; May 02, 2015 [cited 2015 Aug 28]. Available from: http://www.space.com/29295-rocket-history.html

with spaceflight. He wrote in his autobiography about his inspiration as a boy: "I imagined how wonderful it would be to make some device which had even the possibility of ascending to Mars."[3]

As a student at Worcester Polytechnic Institute in Massachusetts, he experimented on a gunpowder-fueled rocket in the basement of the physics building. He attracted enough attention to receive support from his professors. After graduating, Goddard began teaching physics in 1914 at Clark University in Worcester. By 1914, he had already received two patents for his work—one for a liquid-fueled rocket and the other one for a multistage rocket using the liquid fuel as the first stage propellant and solid fuel for the second or third stage. Both were significant milestones for future spaceflight. Goddard continued his theoretical work, predicting the behavior of rockets in the vacuum of space, proposing that air wasn't required for a rocket to produce thrust. In 1919 he published a paper entitled "A Method for Reaching Extreme Altitudes," which became one of the classic texts of twentieth-century rocket science, including the mathematical theories of rocket propulsion.[4]

Dr. Goddard successfully launched the first liquid-fueled rocket (named "Nell") from his Aunt Effie's farm in Auburn, Massachusetts, on March 16, 1926. For greater stability, the heavy rocket motor was placed on top of the rocket with lines to the oxygen and gasoline fuel tanks at the bottom.[5] The rocket reached a maximum of only 41 feet (12.5 m), lasting about two and a half seconds and traveling about 60 mph (96.6 km/hour). This was significant, but not enough to impress government officials to secure funding. In fact, even the press ridiculed his work. After this test flight, a *New York Times* editorial described his work as foolish. Goddard responded to a reporter's question, saying "Every vision is a joke until the first man accomplishes it; once realized, it becomes commonplace (Fig. 6.1)."[6]

A small amount of funding from the Smithsonian Institution, made possible in part by the support of Charles Lindbergh, allowed Goddard to continue his research. In 1930 he moved to New Mexico along with some of his colleagues to continue working away from the public eye. Goddard and his team launched rockets that achieved altitudes up to 1.7 miles (2.7 km)

---

[3] NASA.gov. [Internet] NASA.gov; c2007. Robert Goddard: a man and his rocket; Mar 09, 2004 [cited 2015 Aug 28]. Available from: http://www.nasa.gov/missions/research/f_goddard.html

[4] NASA.gov. [Internet] NASA.gov; c2007. Robert Goddard: a man and his rocket; Mar 09, 2004 [cited 2015 Aug 28]. Available from: http://www.nasa.gov/missions/research/f_goddard.html

[5] NASA.gov. [Internet] NASA.gov; Mar 16, 2001 [cited 2015 Sep 01]. Available from: http://apod.nasa.gov/apod/ap010316.html

[6] NASA.gov. [Internet] NASA.gov; c2007. Robert Goddard: a man and his rocket; Mar 09, 2004 [cited 2015 Aug 28]. Available from: http://www.nasa.gov/missions/research/f_goddard.html

Fig. 6.1 Robert Goddard and "Nell," his first liquid-fueled rocket (Image courtesy of NASA)

with speeds faster than the sound barrier at 550 mph (885 km/hour). He successfully achieved three-axis steerable thrust to the rockets to control the direction of rocket flight. One of his rocket flights in 1929 carried the first scientific payload, a barometer, and a camera. By the time he died in 1945, he was awarded over 200 patents in rocketry although his propulsion research was not widely recognized during his lifetime.[7] Goddard also worked with the US military to create and build the bazooka, an antitank weapon. His theories were expanded on by both German and American scientists to further develop missiles and rockets.[8]

Today Robert Goddard is referred to as the father of modern rocketry, and finally his significant achievements in rocket propulsion have been acknowledged. Unfortunately, he didn't live to see his vision of traveling into space become a reality. One of NASA's facilities, the Goddard Space Flight Center in Greenbelt, Maryland, was named to honor his scientific achievements.

---

[7] NASA Fact Sheet. Goddard Space Flight Center. Robert H. Goddard: American rocket pioneer. FS-2001-03-017-GSFC. 2001.

[8] Lehigh University. [Internet] ei.lehigh.edu; Robert Goddard; [cited 2015 Sep 01]. Available from: http://www.ei.lehigh.edu/learners/energy/readings/people_energy.pdf

Rocket Development During World War II

Rocket development moved into a military phase during World War II. Missile capability became an important military advantage, allowing for the delivery of destructive weapons. The technology required for space exploration turned out to be a hard sell for funding when the world situation demanded a more aggressive approach to weaponry, including weapons of mass destruction. Eventually, the military needs turned out to be the impetus for developing the required space technology. The Germans first adapted Goddard's rocket theories and testing to advance their missile development, including rocket thrust control, all of which were instrumental in the success of the V-2 program.

Following World War II, German rocket scientists working at the rocket development site at Peenemünde emigrated to both the United States and the Soviet Union to eventually contribute to the space race of the 1960s. One of the premier German rocket scientists was Werner Von Braun. His story is chronicled below.

Wernher Von Braun, the Father of Space Travel

Dr. Wernher von Braun (1912–1977) was one of the most important rocket scientists of the modern era and the man who steered the American space program to the Moon. A brilliant engineer, physicist, and inventor, he grew up dreaming of launching rockets into space. As a young man, he was inspired by Herman Oberth's *The Rocket into Interplanetary Space*. When Von Braun studied mechanical engineering and physics in Berlin, he assisted Oberth in his testing of liquid-fueled rockets.[9]

Because of the nature of his work, Von Braun was noticed by the German government, who provided him with research funding. In 1937 he became the technical director of the famous rocket facility located at Peenemünde.[10] Impressed with von Braun and his work, Nazi leadership supported his development of a long-range ballistic missile. The A-4 became the most advanced rocket produced under his leadership. The Nazis changed the A-4's name in 1944 to V-2 (V for *Vergeltung*, the German word for "vengeance"), and they began targeting the rockets to launch toward London and Antwerp. The V-2,

[9] MSFC History office. [Internet] NASA.gov; Recollections of childhood: early experiences in rocketry as told by Werner Von Braun 1963. [cited 2015 Sep 02]. Available from: http://history.msfc.nasa.gov/vonbraun/recollect-childhood.html

[10] National Aviation Hall of Fame. [Internet] nationalaviation.org; c2011. Wernher Von Braun. [cited 2015 Sep 02]. Available from: http://www.nationalaviation.org/von-braun-wernher/

fueled by liquid ethanol, could deliver a 1-ton (900 kg) warhead to a target as far as 200 miles (322 km) away, traveling faster than 3500 mph (5633 km/hour) and reaching altitudes as high as 6 miles (~10 km). The Germans launched more than 1300 V-2s c and hundreds more toward Belgium and France.[11] The V-2 was outfitted with an automatic guidance system that used gyroscopes to track the position of the rocket and adjust the trajectory by changing the position of the rudders attached to the fins without having to control the rocket from the ground.[12]

The V-2s, although successful, were developed too late to make a significant difference in the ending of the war. Following World War II, German rocket scientists working at Peenemünde surrendered and were sent to either the United States or the Soviet Union. Von Braun still had faith in his dream and wrote: "It was the space station we sought and we still seek it wherever we may be. We desire to open the planetary world to mankind."[13] Dr. Von Braun and over 125 key German rocket scientists surrendered to the Americans in 1945 and were transferred to the custody of the US Army. They were sent to Fort Bliss, Texas, where Von Braun was named technical advisor of the White Sands Proving Grounds. His group re-assembled rocket parts from Germany, and in 1946 the V-2 was launched once again, this time with scientific purposes, using instrumentation to study Earth's upper atmosphere.[14]

In 1950, von Braun and his team, including American scientists and engineers were transferred to Huntsville, Alabama. Von Braun, still interested in manned space travel, was thwarted again due to the outbreak of the Korean War. The military focus was again on ballistic missile development. Von Braun directed the army ballistic weapons program and focused his efforts on developing the medium-range Redstone ballistic missile and the intermediate-range Jupiter ballistic missile. Both rockets ended up being vital to the success of the Mercury and Gemini programs. In 1955, von Braun became a US citizen.[15] As a response to the Soviet Union's launch of the satellite Sputnik 4

[11] History.com Staff. [Internet] History.com; c2009. Germany conducts first successful V-2 rocket test; 2009 [cited 2015 Sep 02]. Available from: http://www.history.com/this-day-in-history/germany-conducts-first-successful-v-2-rocket-test

[12] Hollingham, Richard. [Internet] Bbc.com; c2014. V2: the Nazi rocket that launched the space age; Sep 8, 2014 [cited 2015 Sep 02]. Available from: http://www.bbc.com/future/story/20140905-the-nazis-space-age-rocket

[13] National Aviation Hall of Fame. [Internet] nationalaviation.org; c2011. Wernher Von Braun. [cited 2015 Sep 02]. Available from: http://www.nationalaviation.org/von-braun-wernher/

[14] National Aviation Hall of Fame. [Internet] nationalaviation.org; c2011. Wernher Von Braun. [cited 2015 Sep 02]. Available from: http://www.nationalaviation.org/von-braun-wernher/

[15] National Aviation Hall of Fame. [Internet] nationalaviation.org; c2011. Wernher Von Braun. [cited 2015 Sep 02]. Available from: http://www.nationalaviation.org/von-braun-wernher/

Fig. 6.2 Wernher von Braun standing in front of the Saturn V's F1 engines (Image courtesy of NASA)

months earlier, Von Braun's team put Explorer-1, the first American satellite, into Earth orbit in late January 1958. Von Braun's visions and accomplishments became key to helping America enter the space age.[16]

The National Aeronautics and Space Administration (NASA) was formed in 1958. Von Braun became director of the NASA Marshall Space Flight Center and the chief designer of the Saturn V launch vehicle, the rocket that would take Americans to the Moon (Fig. 6.2).[17]

Werhner von Braun became a prominent spokesperson for space exploration in the 1950s through the Apollo program. He was present for the launch of the *Apollo 11* mission using the Saturn V rockets that carried the first astronauts to the Moon. He retired from NASA in 1972. Von Braun stayed active in his conceptual designs, including futuristic permanent wheel-shaped lunar

[16]Wilford, John N. Remembering when U.S. finally (and really) joined the space race. The New York Times. 29 Jan 2008.

[17]History.com Staff. [Internet] History.com; c2010. Von Braun moves to NASA; 2010 [cited 2015 Sep 02]. Available from: http://www.history.com/this-day-in-history/von-braun-moves-to-nasa

space stations that rotated to generate artificial gravity. Von Braun is given credit for the rocket technology that took America to the Moon, winning the space race against the Soviet Union.

The Early History of Space Politics

It is impossible to tell the story of manned spaceflight without a discussion of worldwide politics that accelerated the development of the technology, making space travel possible in what many consider a shortened timeframe. Political priorities determine how government funding is disbursed. In the early days of the space race, being first in the world meant military, scientific, and political supremacy for both the United States and the Soviet Union.

The story of spaceflight started in the 1950s with the political conflict known as the Cold War, which brought the United States to the brink of nuclear war with the communist Soviet Union (USSR). The Soviet Union was an ally to America during World War II, along with Britain and France, when they were united against their common enemy, Nazi Germany. The relationship of the USSR and the Allies began to strain under the rule of Stalin. In 1943, Stalin wanted the Allies to help him set up a second war front because he felt that the Soviet Union was taking on the brunt of the fight against Germany. Once the fighting was over, Stalin thought his country would be weakened, leaving the Allies in a superior position. The Allies disagreed, causing a rift in the relationship. In early 1945, Churchill, Roosevelt, and Stalin agreed to post-war conditions for the countries under Nazi rule and the creation of a UN organization to promote international peace. Stalin's communist viewpoints, not adopted by the Allies, created tensions even as Nazi Germany surrendered. When the United States used atomic bombs to end the war against Japan, Stalin became aware that the Soviet Union was lagging behind in weapons capability despite their overwhelming manpower. By the end of 1945, the Cold War had begun.[18]

Over the next several years, both the United States and the Soviet Union independently developed their military capabilities, which included nuclear bombs and missile delivery systems, much of this work done in secrecy. Tensions increased due to number of provocative events and postures taken by the Soviet Union at this time. Both countries had scientists interested

[18] Historylearning.com Staff. [Internet] Historylearningsite.co.uk; c2015. 1945–1950; [cited 2015 Sep 05]. Available from: http://www.historylearningsite.co.uk/modern-world-history-1918-to-1980/the-cold-war/1945-1950/

and capable of putting a satellite into Earth orbit. Focus on military advantage took priority over scientific space endeavors. As early as the mid-1940s, American proposals to launch a satellite into orbit were discussed but received only mild enthusiasm. In 1946, the US Army Air Corps funded the Douglas Aircraft Company's Project RAND (Research and Development) for a feasibility study on the design and military uses of an Earth orbiting satellite.[19] The resulting report predicted that "The achievement of a satellite craft by the United States would inflame the imagination of mankind, and would probably produce repercussions in the world comparable to the explosion of the atomic bomb."[20] Despite interest expressed by both the US Navy and US Army Air Force in satellite development for military purposes, President Truman had little interest in space systems, preferring aeronautical research. Without presidential interest or focus, the RAND Corporation could only work to further define the military use of satellites and the possibilities of future space exploration. Nothing followed the Project RAND study, leaving the United States lagging behind the Soviet Union who secretly was working on developing an Earth-orbiting satellite and a missile derived launch system.[21]

After the surrender of scientists and engineers at Peenemünde, the Soviet Union ballistic missile program initially took a path similar to the United States. It is generally accepted that the cream of the crop of the engineers ended up in America; however, a large group of Germans went to the USSR to engage in their military and space efforts. A physicist, Helmut Gröttrup, agreed to work with the Soviet Union missile program, hoping to become a leader and separate himself from Von Braun. Remaining in Germany, he began working with colleagues to re-start the production of the V-2, using some of the missile salvage left behind. In 1946, the Soviet government decided to transfer this work back to the USSR, transporting hundreds of scientists and engineers and their families. The Germans would now work with Soviet rocket engineers going forward. Sergei Korolev, who would lead the Soviet space achievements in the 1950s and 1960s, did not welcome the German participation until he learned that the German group's rocket development had actually progressed further than the Russians. Eventually Korolev's plans were abandoned in favor of the development of a V-2 replica.

[19] Kalic, Sean N. US presidents and the militarization of space, 1946–1967. College Station, TX: Texas A&M University Press; 2012. 224 p.

[20] Aeronautics and Space Engineering Board, Division on Engineering and Physical Sciences. Forging the future of space science: the next 60 years. National Research Council; 2010. 166p.

[21] Kalic, Sean N. US presidents and the militarization of space, 1946–1967. College Station, TX: Texas A&M University Press; 2012. 224 p.

The first launches of the remade V-2s occurred in 1947. Often the Soviet engineers shut out the Germans despite their proposed improvements for the V-2 as well as plans for a new guided missile. The Russians found it difficult to give the Germans credit, causing a rift between the two groups of scientists. The Germans were sent to work and live at a remote location and could not collaborate with the Russians or test their concepts. Eventually, they were told that their services were no longer needed and were going to be sent back home to Germany. For most of them, the result was devastating. By 1951, they were allowed to return to East Germany, although the Gröttrups remained until late 1953. Some Germans with particular expertise, such as guidance experts, were transferred to Moscow to work with the Russian scientists.[22]

Korolev received very little support for his satellite work. The Russian focus was now on ballistic missile development and gaining military advantage similar to the United States at that time. Satellite development in both countries would have limped along indefinitely except for international interest among some famous scientists who wanted to study Earth's upper atmosphere. In the early 1950s, American scientists led by James van Allen discussed setting up an international program to study the upper atmosphere and the edge of outer space using sounding rockets, balloons, or ground observations. The U.S. Department of Defense was pursuing research in rocketry and upper atmospheric sciences in order to maintain national leadership in science and technology.

The upcoming period of intense solar activity from July 1957 to the end of December 1958 (named the International Geophysical Year, or IGY) provided the perfect opportunity for the cooperative study of the space environment by scientists of 67 nations. In October 1954, the United States submitted a proposal to orbit an artificial satellite during the IGY. Up to that time, the Soviet participants did not have any submissions but were clearly surprised by the US satellite proposal. Shortly afterward, the Soviets began to look at space exploration with a new and more urgent perspective.[23]

The Soviet Union established a commission in 1955 in response to the American satellite announcement. It stated that "One of the immediate tasks of the Commission is to organize work concerning building an automatic laboratory for scientific research in space."[24] Proposals for scientific experiments that could be mounted on satellites were specifically called for. Although this

[22] Zak, Anatoly. Sep 2003. The rest of the rocket scientists. Air & Space Magazine.

[23] Siddiqi, Asif A. [Internet] History.NASA.gov; c2015. Korolev, Sputnik, and the International Geophysical Year; [cited 2015 Sep 07]. Available from: http://history.nasa.gov/sputnik/siddiqi.html

[24] Siddiqi, Asif A. [Internet] History.NASA.gov; c2015. Korolev, Sputnik, and the International Geophysical Year; [cited 2015 Sep 07]. Available from: http://history.nasa.gov/sputnik/siddiqi.html

group did not work directly with the Soviet missile and space program, it provided a way for Korolev's satellite efforts to be connected with the IGY.

At the end of July 1955, President Eisenhower's office announced that the United States would launch "small Earth-circling satellites" as part of its participation in the IGY. The Soviets soon announced their intention to launch an artificial Earth satellite within 2 years. Both announcements brought a lot of attention worldwide.[25]

Korolev was front and center again for Soviet satellite development. Similar to Von Braun, Korolev was more interested in space exploration than weapons of mass destruction. In the early 1950s, he even wrote a technical report on the possibility of sending a probe to the Moon. See Korolev's story is below.

Sergei Pavlovich Korolev, the Founder of the Soviet Space Program

Sergei Pavlovich Korolev (1907–1966) was the most significant figure in the Soviet space program. In the era following the Bolshevik Revolution in Russia, Sergei became interested in aviation, most likely influenced by his stepfather who was an engineer. In 1924 he attended the Kiev Polytechnic Institute and became involved with gliders as a hobby. Two years later, he transferred to Moscow's Bauman High Technical School, considered to be the best engineering college in Russia. After graduation, he joined the joined the Central Aero and Hydrodynamics Institute and was soon appointed as chief of the Jet Propulsion Research Group, where he led the development of cruise missiles and a manned rocket-powered glider. In 1933, he launched the first liquid-fueled rocket in the USSR and seemed to be rewarded for his hard work and loyalty to the system.[26]

Life in the USSR under Stalin's rule (1929–1953) took on a new dimension. Stalin decided that the use of forced labor would speed up the Soviet Union's modern industrialization and military power from essentially a peasant society. He accomplished this by terrorizing the country, putting millions of people into forced labor camps, called gulags. In mid-1938, Korolev was arrested and sent to a concentration camp in Siberia but later returned in 1940 to a Moscow prison. Due to trumped up charges of sabotage, he was sentenced to 10 years in labor camps. In prison, he was fortunate enough to

[25] Siddiqi, Asif A. [Internet] History.NASA.gov; c2015. Korolev, Sputnik, and the International Geophysical Year; [cited 2015 Sep 07]. Available from: http://history.nasa.gov/sputnik/siddiqi.html

[26] Russian Space Web Staff. [Internet] History.NASA.gov; c2015. Korolev; [cited 2015 Sep 07]. Available from: http://www.russianspaceweb.com/korolev.html

be connected with one of the "design" bureaus, led by Andrei Tupolev. Here Korolev helped develop the Tu-2 bomber, an important aircraft for the Soviet Air Force in World War II. Released from prison in mid-1944, he was made a colonel in the Red Army in 1945 and sent to Germany to evaluate the restoration of the V-2.[27]

Korolev was surprised with the sophisticated design of the V-2's guidance systems and engines. He also knew that some of the best German engineers had gone to the United States with several V-2's intact, along with a lot of plans and salvage parts. Korolev successfully reproduced the V-2 but had his own ideas for a better design. By August 1957, he launched the Soviet R-7 booster, the world's first intercontinental ballistic missile that could travel over 4000 miles. He beat the United States by over a year. Korolev still had a passion for space exploration and a loyalty to a system that treated him so poorly.

In 1957, his dreams became fulfilled with the successful launch of Sputnik 1 on top of a modified R-7 booster. He was then instrumental in putting into space the first dog, the first two-man crew, the first woman, and the first three-man crew. He directed the first walk in space, created the first Soviet spy satellite and communication satellite, built launch vehicles, and flew spacecraft towards the Moon, Venus, and Mars. Finally, the launch of the first man into space (Cosmonaut Yuri Gagarin) ensured Korolev his place in history. His family of R-7 space boosters launched Russian cosmonauts into orbit for decades (Fig. 6.3).[28]

Sergei Korolev displayed incredible intelligence, energy, and an unwavering belief in spaceflight. He turned the weapons of the Soviet Union into peaceful instruments of space exploration, forever labeling his country as the first nation to travel in space. Before his death in January 1966, he was developing the N1 Moon rocket while under political pressures much the same as Von Braun's team. Korolev's accomplishments were only recognized by Soviet authorities after his death.[29]

Korolev proposed modifying his original R-7 booster to launch a series of satellites into Earth orbit. He had a specific timetable in mind, which he secretly shared with his colleagues. He had hoped to begin the first launches in April–July 1957, before the start of the International Geophysical Year. The official go-ahead on the project had not yet been issued, and there still existed mixed feelings on what was perceived as a civilian effort as compared

[27] Russian Space Web Staff. [Internet] History.NASA.gov; c2015. Korolev; [cited 2015 Sep 07]. Available from: http://www.russianspaceweb.com/korolev.html

[28] Flashback to sputnik launch. 2007. Irish Times; 15.

[29] Rodgers P. 2011. The man who fell to earth. The Independent on Sunday; 18.

Fig. 6.3 Sergei Korolev (on the right) with Yuri Gagarin, the first human to fly in space, taken at the Gargarin Museum, Star City, Russia (Image courtesy of NASA)

to military operations. Approved as a scientific endeavor, this was still not considered a top priority.

Korolev decided to launch a simpler and lighter satellite into orbit instead of a more sophisticated scientific laboratory that weighed more than a half of a ton. Korolev proposed two small satellites, each with a mass of 40–50 kg, to be launched in the period of time immediately prior to the IGY. However, the first three launches were all failures, and the pressure on Korolev and his team was immense. It was known that the United States planned to launch an artificial satellite under the name Vanguard, using a three-stage missile developed by the Naval Research Laboratory. In September 1956, the first American attempt to launch a satellite failed. Several additional test launches were scheduled, and it was felt that there was still time to meet the IGY deadline. Korolev successfully launched the next two R-7's in August and September of 1957, and the stage was set for the launch of Sputnik.[30]

[30] Siddiqi, Asif A. [Internet] History.NASA.gov; c2015. Korolev, Sputnik, and the International Geophysical Year; [cited 2015 Sep 07]. Available from: http://history.nasa.gov/sputnik/siddiqi.html

The Space Race Heats Up

Political posturing had continued between the United States and the USSR through the mid-1950s. Nikita Khrushchev, a rising political star in the Communist Party, was looked on with favor by Stalin and held a number of high ranking positions in the party, including First Secretary of the USSR in 1953. After Stalin's death in 1953, Khrushchev built up his political support and used his influence to get his nominee (Nikolay Bulganin) elected as premier in 1955. Khrushchev, seen as the man with the real power, eventually became premier himself in 1958.[31] The Soviet Union hoped to achieve communist domination of the world and was willing to take aggressive steps to achieve that goal, short of war. Because both the United States and the Soviet Union had nuclear capability and the means to deliver it, Khrushchev decided that the best choice would be peaceful coexistence. Other than military achievements, both countries became aware that advancements in science and technology and space exploration would be a clear demonstration of the superiority and capabilities of each nation.

As the end of 1957 approached, neither nation had achieved success in launching an artificial satellite. But then, before a third Vanguard test launch could occur, on October 4, 1957, the Soviet Union launched the world's first artificial satellite, Sputnik I. Upon hearing the news, the rest of the world was shocked, surprised, in disbelief, or dismissive of the importance of the event. Americans felt that they were dealt a psychological blow. Scientists were devastated that they were one step behind the Soviets in achieving an event of this significance. In addition, the Soviet capability of putting a satellite into orbit also meant they had the capability of launching ballistic missiles carrying nuclear or conventional weapons to attack other countries. The world changed forever. The satellite's signal, a recognizable beep-beep-beep, heard around the world, reminded Americas that they could no longer claim their capitalist system was superior to the communist Soviet Union.[32] Before long, in early November, the Soviets launched Sputnik II, carrying a heavier payload that included a dog named Laika.

US government officials evaluated Sputnik in light of past political policy and changes made going forward. The Eisenhower Administration found itself criticized for not focusing on the importance of space and dismissing its

[31] Global Security Staff. [Internet] globalsecurity.org; c2015. 1955–1964—Kruschev; [cited 2015 Sep 08]. Available from: http://www.globalsecurity.org/military/world/russia/khrushchev.htm

[32] Chalmers M. Roberts, Staff Reporter. 1957. Sputnik healthily destroyed some illusions. The Washington Post and Times Herald (1954–1959); 1.

significance in lieu of military advancements. President Eisenhower thought that Sputnik was a trick of sorts, without any real substance to follow. Decades later, some analysis reveals that prior to Sputnik, Eisenhower's focus was on missile and military satellite technologies that he thought would contribute to national security in the future. After Sputnik, Eisenhower was forced to re-examine his approaches to achieve American superiority. The fact that the United States, at least in the eyes of Americans, now trailed the Soviet Union was an image that was difficult to confront and would challenge Eisenhower's leadership.[33]

The US Defense Department awarded funding for a simultaneous project to Vanguard and to Werhner von Braun's Army Redstone Arsenal team to begin work on the Explorer project. In a very short period of time, by the end of January 1958, the United States successfully launched Explorer I, a rocket carrying a satellite with a small scientific payload to measure magnetic radiation in space. The Explorer program continued on with a series of scientific experiments. On October 1, 1958, Congress created the National Aeronautics and Space Administration (NASA) from the National Advisory Committee for Aeronautics (NACA) and other government agencies.[34]

The governments of both the United States and the Soviet Union were now focused and invested in space-related missions. Scientific and space achievements were happening on an accelerated timeline. President Eisenhower was interested in pursuing cooperative space initiatives with the Soviets, thinking that this alliance would ensure space would be used for peaceful purposes. Khrushchev, not interested, instead made military demands. The United States pursued a legal framework for peaceful space activities that eventually led to the Outer Space Treaty of 1967. The Soviet space program expressed an interest in sending probes toward the Moon. In January 1959 *Luna 1* became the first manmade object to orbit the Sun, leading to *Luna 2,* which impacted the Moon in September and *Luna 3,* which orbited the Moon and photographed the surface in October. The US *Pioneer 4* had a successful Moon flyby in March 1959. Eisenhower's viewpoint remained unchanged—he thought that space projects didn't add any value to the nation's security and that they were costly without obvious reward. He supported scientific experimentation with specific objectives instead of voyages of exploration.[35]

---

[33] Mieczkowski, Yanek. Eisenhower's Sputnik moment: the race for space and world prestige. Ithaca, New York: Cornell University Press; 2013. 368p.

[34] Coldwar Org Staff. [Internet] coldwar.org; c2015. Sputnik; [cited 2015 Sep 08]. Available from: http://www.coldwar.org/articles/50s/sputnik.asp

[35] Planetary Staff. [Internet] planetary.org; c2015. Sputnik; [cited 2015 Sep 08]. Available from: http://www.planetary.org/explore/space-topics/space-missions/missions-to-the-moon.html#pioneerp3

Soon after the creation of NASA in October of 1958, several organizational changes were made to focus more on a manned space program called Project Mercury, which would test a person's ability to survive and work in space. NASA's long-term planning for the Moon missions was being done simultaneously along with Project Mercury. Government reorganization resulted in the military efforts of Pioneer and Vanguard as well as the Army Ballistic Missile Agency, with its Saturn rocket programs being placed under the guidance of NASA. Military missile and space rocket development efforts would be more efficient if integrated.

Over 50 years later, it is widely thought that support for the space program in the late 1950s was widespread, when in fact many scientists actually opposed the fast and furious approach that the space race was requiring. A 1958 report to the president from his Science Advisory Committee stated that some of the most distinguished scientists in America were more interested in the importance of science related to Earth rather than space ambitions. The fear that pursuing space science might weaken scientific efforts in other areas promoted a more balanced effort between science and technology. Because of the escalation of events by the Soviets, the balanced science budget never happened.[36]

In April, NASA selected seven astronauts for the manned space program called Mercury. Meanwhile, Korolev was launching a series of Vostok spacecraft capable of launching the first humans into orbit by 1961.[37]

John F. Kennedy defeated Richard Nixon, Eisenhower's vice president, in the presidential election of November 1960. Despite Eisenhower's lack of enthusiasm for costly space efforts, the organizational structure under NASA, already in place when Kennedy took office, made the upcoming accelerated race to the Moon go much smoother.

The Cold War and the space race led to a lot of secrecy on both sides. Some of Eisenhower's efforts for cooperative military and space efforts were not achieved because of political events, such as the downing of a U-2 spy plane over the Soviet Union in May 1960. President Kennedy also tried to achieve space cooperation with the Soviets. In his inaugural speech, he said "Let both sides seek to invoke the wonders of science instead of its terrors. Together let us explore the stars."[38]

---

[36] Madrigal, Alexis. [Internet]. theatlantic.com; c2015. Moondoggle: the forgotten opposition to the Apollo program. Sep 12, 2012. The Atlantic—Technology. [cited 2015 Sep 09]. Available from: http://www.theatlantic.com/technology/archive/2012/09/moondoggle-the-forgotten-opposition-to-the-apollo-program/262254/

[37] Au.af.mil Staff. [Internet] Au.af.mil; c2015. Eisenhower years: 1953–1960; [cited 2015 Sep 09]. Available from: http://www.au.af.mil/au/awc/awcgate/au-18/au18003c.htm

[38] Sagdeev, Roald & Eisenhower, Susan. [Internet] NASA.gov; c2008; United States-Soviet space cooperation during the Cold War. [cited 2015 Sep 09]. Available from: http://www.nasa.gov/50th/50th_magazine/coldWarCoOp.html

The Soviet Union had no interest in cooperation at first, only in achieving firsts. On April 12, 1961, the cosmonaut Yuri Gagarin became the first human to be launched into space and orbit Earth. Alan Shepard followed with his Mercury mission on May 2. After America's John Glenn's orbiting flight, Khrushchev indicated some willingness to cooperate, an agreement that eventually led to the exchange of weather data from satellites, coordinated launches of meteorological satellites, a joint scientific effort to map the geomagnetic field of Earth, and cooperation in the relay of communications data.[39]

The American space program continued one step behind for several years, missing out on a considerable number of firsts after the first human in space—first human in orbit, first animal in orbit, first woman in space, multi-person mission, first commercially used satellite, first simultaneous flight of multi-person spacecraft, first extravehicular activity, first probe to orbit the Moon, first probe to land on the Moon, first automated (crewless) rendezvous and docking, first docking between crewed spacecraft, and so on. The Soviets wanted all of the firsts that they could accumulate on their way to landing on the Moon. The Soviet secret program hid the number of accidents and lives lost in their pursuit.[40]

Shortly after Alan Shepard's flight, in his first State of the Union address, President John F. Kennedy accelerated the space program, setting a goal for the United States to land a man on the Moon before the end of the decade. He was motivated to inspire Americans to catch up and overtake the Soviets in the space race. A political embarrassment from the Bay of Pigs disaster in mid-April and a sense of urgency to prevent communism from spreading further motivated Kennedy. The events leading to the landing on the Moon cannot be examined and understood without looking through the lens of the Cold War.

Kennedy's speech set the stage for the challenge: "I believe this nation should commit itself to achieving the goal, before this decade is out, of landing a man on the Moon and returning him safely to Earth. No single space project in this period will be more impressive to mankind, or more important in the long-range exploration of space; and none will be so difficult or expensive to accomplish."[41]

---

[39] Sagdeev, Roald & Eisenhower, Susan. [Internet] NASA.gov; c2008; United States-Soviet space cooperation during the Cold War. [cited 2015 Sep 09]. Available from: http://www.nasa.gov/50th/50th_magazine/coldWarCoOp.html

[40] Braeunig, Robert. [Internet] Braeunig.us; c2011. Manned space flights; [cited 2015 Sep 09]. Available from: http://www.braeunig.us/space/manned.htm

[41] NASA.gov. [Internet] NASA.gov; c2014. NASA—excerpt from the 'special message to the congress on urgent national needs'; [cited 2015 Sep 09]. Available from: https://www.nasa.gov/vision/space/features/jfk_speech_text.html#.VfDorxFVhBc

Kennedy's bold political statement to land a man on the Moon and return him to Earth in the 1960s decade surprised even NASA and all involved in space exploration. Funding and resources for this endeavor would not be issues, but it would be a huge and risky undertaking, the likes of which were not experienced since the Manhattan Project during World War II, when the atom bomb was developed. The Soviets were put on notice.

NASA's efforts from that day forward would continue to use Kennedy's speech as a guide. Project Mercury would blend into the multi-crew Gemini program, and finally the Apollo program would land a crew on the Moon and bring them safely back to Earth. During this time, tensions between the United States and the Soviet Union would increase, due to significant world events, including the construction of the Berlin Wall in 1961, the Cuban Missile Crisis of 1962, and the outbreak of the war in Southeast Asia. Both countries moved toward nuclear war on more than one occasion. The political events became the backdrop for the all-important race to the Moon. A sense of urgency existed to fulfill the Kennedy dream, especially after his assassination in November 1963. Lyndon Johnson, sworn in as president, took over the reins and worked toward Kennedy's Moon landing goal. In January 1967 in a routine ground test in the newly developed Apollo capsule, the crew of *Apollo 1* died strapped into their seats on top of the Saturn V rocket. The program halted to investigate the causes of the accident and the successful landing on the Moon before the end of the decade was very much in jeopardy.

Krushchev led the Soviet Union as its premier until 1964. He was a complex political figure, who in the end chose peaceful coexistence with America over nuclear war. He was responsible for provocative behavior such as putting nuclear weapons just off of Florida coast in the Cuban Missile Crisis and constructing the Berlin Wall, which separated East and West Berlin.

The end of the space race occurred on July 20, 1969, when *Apollo 11* commander Neil Armstrong stepped onto the Moon's surface. The feeling in the Soviet Union most likely mirrored the American feeling when Yuri Gagarin became the first man in space. Information in the Soviet Union was not easily disseminated, and even if it was, there was a significant amount of propaganda associated with the news. National pride was at a high in America.

Final Thoughts on the Politics of the Space Race

Because the space race and the Cold War were so intertwined, it is difficult to analyze the politics of space exploration without involving the possibility of a nuclear war at that time. Fear of war became intertwined with the hopes of landing on the Moon before the enemy. Fortunately, cooler heads prevailed on both sides of the conflict, and we all lived to see another day. We had advanced outside of Earth's atmosphere and landed on the Moon, achieving an amazing scientific and technological achievement. It was not without cost. Three astronauts lost their lives in the Apollo spacecraft testing. It is not known how many cosmonauts lost their lives at that time because their program was shrouded in so much secrecy. The Soyuz 1 capsule crashed into Russian soil in 1967, killing the first man in space. Similar to the problems with the *Apollo 1* capsule, sources say the Russians knew that the capsule would fail, but leadership ignored the warnings. A Soviet launch pad explosion in 1960 killed over 125 people in a horrific fire.[42]

When we look back on the space race, it is hard not to think about the contributions made by the German Peenemünde scientists and engineers involved in both the American and the Soviet space programs. The Germans didn't design Sputnik or its rocket, but their ideas influenced the designs and speeded up the effort. Wernher von Braun and his team answered Sputnik with the satellite Explorer 1 and the development of rocket launch systems including the Saturn V. Clearly, they were leaders in the field of missiles and rocketry.

Finally, politics clearly played a role in the space race. Never was there a time that more clearly illustrated how space research and exploration were interconnected with political events and how space was representative of military power and scientific achievement. After the *Apollo 11* Moon landing, the tension dissipated, and a new era in space exploration began.

[42] Smolchenko, Anna. [Internet] phys.org; c2010. Russia marks 50 years since horrific space launch disaster; Oct 24, 2010; [cited 2015 Sep 09]. Available from: http://phys.org/news/2010-10-russia-years-space-disaster.html

7

The Post-Apollo and Space Shuttle Era

Keywords Apollo • Christopher C. Kraft • Robert Crippen • Defense Advanced Research Projects Agency (DARPA) • Department of Defense • European Space Agency (ESA) • Max Faget • James Fletcher • Hubble Space Telescope • Manned Orbiting Laboratory (MOL) • Marshall Space Center • Michael Griffin • Saturn V • Star Clipper • International Space Station (ISS) • Richard Nixon • Single-stage-to-orbit (SSTO) vehicle • Skylab • Space Doctrine • Spacelab • Space Shuttle • Space Station • Space Task Force • Vandenberg Air Force Base • Wernher von Braun • X-30

> "And as we know now, and as I pointed out many times, the great plume of fire at the bottom of the space shuttle is actually dollar bills burning, and the most efficient method of destroying American dollar bills as has ever been devised by man."
>
> –Representative Dana Rohrabacher, Chairman of the Subcommittee on Space and Aeronautics, during fiscal year 1998, NASA authorization hearings, March 4, 1997.

The national mood in the United States after Neil Armstrong walked on the Moon was one of relief and pride. America had accomplished what had seemed impossible. The promise of safely landing a man on the Moon by the end of the 1960s had been realized, despite President Kennedy's assassination. The United States had beaten the Soviet Union to the Moon and somehow the world felt a lot safer.

L. Dawson, *The Politics and Perils of Space Exploration*,
Springer Praxis Books, DOI 10.1007/978-3-319-38813-7_7

Lives were lost on both sides (both American and Soviet space programs), and yet all of the sacrifices made contributed to the higher purpose of landing and walking on the Moon and returning safely. But almost as quickly as the anticipation and excitement that had built up to the Moon landing, it dissipated. People went on with their daily lives. Ratings for broadcasts for launches and Moon walks after *Apollo 11* were so low that they were taken off of the air. Oddly, the only mission that drew any interest was *Apollo 13*, but only after its in-flight accident and the near deaths of the astronauts.

The average public was not in tune with the specifics of scientific experiments conducted on the surface of the Moon. Each moon landing, no matter how adventurous or dangerous, eventually became commonplace. In fact, with no evidence of life on the Moon, there was little interest in knowing more about the Moon and how this knowledge could benefit humans on Earth.

This chapter examines the politics of space exploration both nationally and internationally in the years following the space race up through the space shuttle era and the development of the International Space Station.

Post-Apollo Space Exploration Politics

Apollo 11 was a huge success, both scientifically and politically. NASA's original plans included lunar missions through *Apollo 20. Apollo 13* ended prematurely due to an almost fatal explosion that challenged the technology and NASA engineers to solve a life-threatening situation. The public became interested in the survival of the crew, but once the astronauts returned safely back to Earth, life returned to normal, focused on Earth-related events. Occasionally, there would be a televised Moonwalk, showing off the rover capabilities to transport the astronauts away from the lunar lander to perform experiments and collect samples. But, as exciting as these ventures were to the scientists and engineers, the general public and the government were focused on other priorities—the Vietnam War and domestic issues associated with diversity and equality.

John F. Kennedy's Moon landing promise was accomplished during the administration of Richard Nixon, who succeeded President Johnson in January 1969. By 1970, NASA was already looking to it post-Apollo objectives, which included the Skylab space station and a reusable shuttle. *Apollo 20* was canceled to save a Saturn V rocket assembly that would be used to carry a Skylab module into Earth orbit. The three final Moon landings were delayed past 1973-4 in order for Skylab to be launched. Because *Apollo 13* never landed on the Moon, its mission activities were reassigned to *Apollo 14*.

NASA decided that *Apollo 17* would be the last Moon flight, marking the end of an era. President Nixon expressed interest in canceling Apollo 16 and 17 in order to move more quickly on to Skylab and the space shuttle. Fortunately, that didn't happen. However, the three canceled missions could have revealed more about the Moon, landing at three additional locations and performing important scientific experiments.[1]

In 1975, there was one final Apollo mission. NASA launched the Apollo part of the Apollo-Soyuz Test Project, a US-Soviet demonstration of peace and cooperation between the two nations. It was the first spaceflight that included two countries working together using their own national spacecraft. The Apollo command module rendezvoused with a Soyuz spacecraft. The event featured symbolic handshakes between three astronauts and two cosmonauts. This powerful gesture indicated that the space race was over and that the two nations could work together on future collaborations in space. However, the reality was that the Skylab mission occurred in Earth orbit and was unrelated to a Moon mission.[2]

The primary reasons for the Apollo cuts were budgetary restrictions imposed by Congress and the Nixon Administration. NASA's budget had risen to an all-time high in the 1960s, but after the primary goal of landing on the Moon was achieved, there was no longer the same urgency to continue travel to the Moon. The relationship between the United States and Soviet Union had settled down, and nuclear war had been averted. Just after the Moon landing in July 1969, NASA's workforce dropped from 400,000 employees and contractors to less than 200,000 in just 6 months. NASA had plans for more cuts as well.[3] There was a declining interest in lunar missions along with a desire to move on to other space exploration goals rather than continue to fund additional manned missions the Moon that carried a high amount of risk. *Apollo 13* had reminded everyone of the dangers of spaceflight.

NASA's twin priorities of developing a manned orbital laboratory and a reusable space vehicle were competing for resources. Wernher von Braun became concerned about the direction NASA was taking. "The legacy of Apollo has spoiled the people at NASA. They believe that we are entitled to this kind of a thing forever, which I gravely doubt. I believe that there may

[1] Silber, Kenneth. 16 July 2009. Down to Earth: the Apollo Moon missions that never were. Scientific American.

[2] Howell, Elizabeth. [Internet] Space.com; c2013. Apollo-Soyuz test project: Russians, Americans meet in space; April 25, 2013 [cited 2015 Sep 21]. Available from: http://www.space.com/20833-apollo-soyuz.html

[3] Silber, Kenneth. 16 July 2009. Down to Earth: the Apollo Moon missions that never were. Scientific American.

be too many people in NASA who at the moment are waiting for a miracle, just waiting for another man on a white horse to come and offer us another plant, like President Kennedy."[4] Von Braun felt that the Apollo program exhausted all of the resources, and when the vision was gone, the feeling was to quit.

Shortly after Richard Nixon became president (February 1969) he established a Space Task Force, an ad hoc committee formed to examine the future of space exploration and outline a number of objectives and priorities. This committee was historically confused with the Space Task Group formed in 1958 to manage manned spaceflight programs. President Nixon was interested in developing some options for a post-Apollo space program that could be associated with his legacy in a similar way to President Kennedy and the Moon landing. The recommendations were positive in their approach: to continue space exploration for peaceful purposes and to include manned space missions as an important component for certain scientific endeavors. The Task Force thought that a strong personal identification for manned space efforts still existed, and, in order to capture the public interest again, it was important to continue these efforts in addition to unmanned missions.

The Task Group concluded there could be a successful manned Mars mission in 15 years, but it would require a total focus of funds at the expense of other objectives. The recommendation was to land a crew on Mars by the end of the twentieth century. This timeframe came and went, with only unmanned missions and probes sent to Mars. Other recommendations included unmanned and a select few manned programs to advance science and engineering and international relations for Earth's benefit, the enhancement of national security without provoking other nations, the achievement of scientific and technological returns from space investments that included a better understanding of the universe, and the development of a low-cost, reliable, reusable, operational space station. This space station could be used for a long time in the future to study Earth and outer space and to supply smaller modular space stations. International cooperation and interests would be part of this last objective. Recommendations seemed to overlap government policy goals and objectives for the future of space flight.[5]

President Nixon overall responded favorably to the recommendations of the report. He outlined six main objectives for NASA going forward.

---

[4] Neufeld, Michael J. Von Braun; dreamer of space, engineer of war. UK: Vintage; 2012. 624 p.

[5] Space Task Group (US). The Post-Apollo space program: directions for the future. [Internet] History Office, NASA Headquarters, Washington, DC; 1969 [cited 2015 Sep 21]. Available from: http://www.hq.nasa.gov/office/pao/History/taskgrp.html

- Continue to explore the Moon, maximizing scientific discovery and ensuring the safety of the crews. Decisions on future missions would depend on the results of previous missions.
- Launch a series of satellites in Earth orbit to study Earth, the universe, and the Solar System as well as sending unmanned spacecraft to all planets of the Solar System, including landing an unmanned vehicle on Mars. The major long-range goal in this objective was to send humans to explore Mars.
- Reduce the cost of space exploration and operations by developing less expensive multi-use, reusable transportation such as space shuttles.
- Expand the study of the human capability to live and work in space for extended periods of time. This would be accomplished in part through a large Earth orbiting laboratory called the Experimental Space Station. The major long-range goal in this objective would be to develop a multi-purpose platform that would serve the purpose of a steppingstone for interplanetary flight.
- Expand the applications of space technology such as using satellite platforms to assess Earth's environment and resources. In addition, develop space-related technology to include applications of meteorology, navigation, national defense, and communications.
- Encourage extensive international cooperation in space in order to achieve progress faster and contribute resources for the benefits of multiple nations.[6]

Nixon's overriding message, however, was to cut back on large-scale space projects. Shortly after Nixon's response, *Apollos 18* through *20* were canceled. The only remaining Apollo-related mission that survived was Skylab, the space station that hosted three crews of three astronauts. Nixon ended up endorsing the space shuttle project. There was resistance in Congress to fund expensive space projects, but the president rightly concluded that the United States could not afford to *not* have a space program. Ending manned spaceflight entirely would have destroyed America's aerospace industry. The space shuttle provided a perfect balance of manned scientific missions that would provide a practical benefit for the space program. The cost of space travel was projected to be lower by using a mostly reusable spacecraft. Future space missions, such as a space station, would become more feasible with payloads transported by a space shuttle. The space shuttle was barely approved by Congress in 1972.

[6] Whittington, Mark. [Internet] Examiner.com. The 1969 space task group and why it failed to chart a post-Apollo space program. [cited 2016 Feb 22]. Available from: http://www.examiner.com/article/the-1969-space-task-group-and-why-it-failed-to-chart-a-post-apollo-space-program

NASA had lobbied for an aggressive approach to maintain superiority in space. That included the development of a reusable space shuttle, a permanently manned space station, and a manned mission to Mars. The ultimate decisions made for the future of space exploration essentially followed a complex series of prioritizations made by President Nixon. Nixon decided that the space program should be considered on the same level as other domestic needs with no privilege. He also decided not to pursue human spaceflight beyond low Earth orbit because of the required financial investment and a lack of desire to revisit the funding levels of the Apollo years. No other lofty human spaceflight goal such as landing a human on Mars should be pursued for the foreseeable future. The primary post-Apollo NASA program would focus on the development of the space shuttle. However, the specific goals and long-term strategy for the shuttle was unclear. During the Nixon Administration, NASA was categorized as one of many domestic programs, and NASA's manned space program was confined to below low Earth orbit, restricting more ambitious missions. There was increased international participation in manned programs, although this was primarily seen by Nixon as foreign astronauts flying on US spacecraft. NASA had additional interests in technology and hardware contributions from other nations.[7]

It is thought by some that President Nixon's decisions concerning the future of space exploration had a greater impact on NASA than John F. Kennedy's vision to land on the Moon. Nixon's "space doctrine," illustrated in a speech by Nixon in March 1970, detailed the policies that have remained at the core of US space policy to this day. His thoughts on the future of NASA:

> We must think of [space activities] as part of a continuing process… and not as a series of separate leaps, each requiring a massive concentration of energy. Space expenditures must take their proper place within a rigorous system of national priorities.…What we do in space from here on in must become a normal and regular part of our national life and must therefore be planned in conjunction with all of the other undertakings which are important to us.[8]

Nixon's focus was to pull back overspending and fund low-cost research efforts and space projects. In addition, NASA funding would compete against other national programs for government resources. In 1972, Nixon met with Dr. James Fletcher, NASA administrator, and shortly afterward announced

[7] Logsdon, John M. After Apollo? Richard Nixon and the American space program. NY: Palgrave Macmillan; 2015. 368 p.

[8] Barber, Chris. [Internet] Nixonlegacy.org. The dawn of the space shuttle. [cited 2016 Feb 22]. Available from: http://www.nixonlegacy.org/the-new-nixon/2016/1/dawn-space-shuttle

the funding of the shuttle program, setting the path for the future of space exploration. President Nixon described the possibilities that the space shuttle would bring:

> "Economy in space will be further served by the space shuttle, which is presently under development. It will enable us to ferry space research hardware into orbit without requiring the full expenditure of a launch vehicle, as is necessary today. It will permit us to place that hardware in space accurately, and to service or retrieve it when necessary instead of simply writing it off in the event it malfunctions or fails. In addition, the shuttle will provide such routine access to space that for the first time personnel other than trained astronauts will be able to participate and contribute in space as will nations once excluded for economic reasons".[9]

Once again, it was a series of presidential decisions that guided the future of the US space program. This time, these decisions were focused on the development of the space shuttle. After the Apollo program, there were no compelling arguments for big manned missions, no political or pressing scientific reasons for a human to go to Mars. Scientists were satisfied sending probes and rovers to discover things about the Martian surface.

President Nixon did not treat the space program as a pressing special interest need. Lack of vision guided the space program for decades to come. Space ambitions for human travel outside of Earth's orbit and to Mars were shelved for another 40+ years due to insufficient resources and focus to achieve those goals. In addition, public opinion was lacking and maybe uninformed. If there were strong reasons for a human Mars mission, they were not lobbied in order to gain national support. In the absence of that, the future was guided by presidential leadership.[10]

Space Stations

Although the decision for manned lunar missions was expensive and all-encompassing for NASA, there was still room left to discuss future efforts, one of which was a space station. The original vision, however, would have to be modified, be less complex and smaller in order to be affordable. In the early 1960s, at Marshall Space Flight Center, Wernher von Braun lobbied for

[9] Logsdon, John M. After Apollo? Richard Nixon and the American space program. NY: Palgrave Macmillan; 2015. 368 p.

[10] Logsdon, John M. After Apollo? Richard Nixon and the American space program. NY: Palgrave Macmillan; 2015. 368 p.

Fig. 7.1 Early concept of Von Braun's envisioned rotating wheel space station (Image courtesy of Marshall Space Flight Center, NASA)

a large Earth orbiting station from which spacecraft traveling to the Moon could be launched, with surrounding orbital tankers that could fuel the vehicles (Fig. 7.1).

In mid-1962, NASA decided that Apollo would use a lunar orbit rendezvous, cutting off Earth orbiting operations. From that time forward, space stations were relegated for future studies to identify their place in the space program. NASA was so heavily invested in the Apollo program that most of this futuristic work paid by the government ($70 million for over 140 contracts) was allocated to aerospace contractors (12 companies).[11] The contracts dealt with all types of future missions, both manned and unmanned, as well as the requirements for rockets and the spacecraft required to carry out the experiments in low Earth orbit, as well as explorations to other planets. It was thought that this was an appropriate time to think about the future of space exploration beyond reaching the Moon.

A variety of space station designs were proposed, some of them big and complex, generating artificial gravity by rotating the station. Von Braun was an advocate of the rotating station that could be launched on the Saturn V

[11] Compton, W. David and Benson, Charles D. The NASA History Series. [Internet] history.nasa.gov; SP-4208 Living and working in space: a history of Skylab; [cited 2015 Sep 22]. Available from: http://history.nasa.gov/SP-4208/ch1.htm

and assembled in space within 4 years. Von Braun's team at Marshall Space Flight Center proposed that the station's purpose be an Earth orbiting base for manned flights to Mars. NASA defined the purpose and plans for an Earth orbiting laboratory by the middle of 1963. The early missions would study humans working and living in space for long periods of time under weightless conditions. An on-board manned laboratory would be used for research and experiments that would look back toward Earth and out into space. To the extent possible, vehicles and equipment from Gemini and Apollo would be used to outfit the station.[12]

To Von Braun's disappointment, the proposals for large and/or rotating space stations were determined to be too expensive, and there were no convincing reasons for a station that big. Von Braun was outspoken about NASA's dependence on the support of Congress. He expressed concern that NASA was less interested in planning for long-term space exploration and more interested in doing what fit into smaller objectives without specific long-range goals. As funding changed, NASA had to respond with changes in future plans. As NASA's budget was reduced in 1964, the future of a medium or larger size space station was in jeopardy. The vision would have to be restricted to an extension of the Apollo program as an Earth-orbiting laboratory. At this time, the air force also began showing interest in an orbiting laboratory that would be viewed as Skylab's competitor.[13]

The US Air Force vs. NASA

The US Air Force showed interest in the construction of a Orbiting Laboratory (MOL), a top secret project, in the mid to late 1960s. The Dyna Soar project had been canceled in 1963, and the Air Force, eager to participate in the development of a space station laboratory, also proposed that up to four military crewmembers work and stay in a space module for extended periods of time working on military related activities. The orbiting laboratory's use was outwardly proposed as scientific research, but the agenda was clearly to provide a reconnaissance window to Soviet activities as well as to study military applications in space. The pressurized module the size of a van would be

[12] Compton, W. David and Benson, Charles D. The NASA History Series. [Internet] history.nasa.gov; SP-4208 Living and working in space: a history of Skylab; [cited 2015 Sep 22]. Available from: http://history.nasa.gov/SP-4208/ch1.htm

[13] Compton, W. David and Benson, Charles D. The NASA History Series. [Internet] history.nasa.gov; SP-4208 Living and working in space: a history of Skylab; 1983 [cited 2015 Sep 22]. Available from: http://history.nasa.gov/SP-4208/ch1.htm

connected to a modified Gemini capsule and launched into low Earth orbit using a military Titan III rocket.[14]

The Air Force was willing for NASA to be responsible for the design and development work needed to fly the military astronauts to the MOL. NASA pushed back, citing the fact that they were a civilian agency with the goal of the peaceful uses of space. The compromise in early 1964 stated that NASA would provide support but the Air Force would develop most of the project. There were three modules planned—for earth science study, astronomy, and testing space systems such as solar panels. A Gemini capsule attached to the module would be used for the astronauts to return to Earth. Problems existed for the MOL. It had an equatorial orbit that would keep it from passing over the Soviet Union that eliminated the reconnaissance part of the mission unless orbital changes were made. Studies were conducted on how to optimize the MOL for NASA as well as military objectives. As the MOL was gaining momentum, the cost of President Johnson's domestic programs as well as the Vietnam War took priority in the budget. The Air Force 1968 budget to develop the station was slashed.

After Nixon was elected president at the end of 1968, the budget numbers became worse for the MOL. The United States went into a recession, and the amount of available money dropped again. Secretary of Defense McNamara, a strong supporter of the MOL, was replaced. In addition, the war continued to use up the military budget, so future space-related programs couldn't survive. The new Secretary of Defense Laird canceled the MOL program altogether. NASA administrator, Tom Paine, was now supporting a new spaceplane to follow the Apollo program. His plan was approved under the condition that the MOL's missions could be combined somehow with the space shuttle's plans.[15]

Skylab

In the early 1960s, NASA was planning a smaller, more efficient space station in low-Earth orbit that could utilize surplus rockets and space hardware from Gemini and Apollo missions. The "Apollo Applications Project" would use

[14] Dorr, Robert F. [Internet]. Defensemedianetwork.com; Air Force Manned Orbiting Laboratory (MOL) astronauts would have conducted surveillance and scientific research; Oct 19, 2011 [cited 2015 Sep 23]. Available from: http://www.defensemedianetwork.com/stories/what-might-have-been-manned-orbiting-laboratory-mol/

[15] Wordpress Staff. [Internet] wordpress.com; False steps: the Space Race as it might have been; the manned orbiting laboratory: a USAF space station; July 15, 2012 [cited 2015 Sep 24]. Available from: https://falsesteps.wordpress.com/2012/07/15/the-manned-orbiting-laboratory-a-usaf-space-station/

a hollowed Saturn V upper stage as a small module that Apollo command modules could dock with. The space station could be used for studying Earth while testing the long-term effects of space travel. The station would be called Skylab. In mid-1965, the Skylab program office was established by NASA headquarters. For the next 4 years, the design evolved.

Skylab took a back seat to Apollo until at least the first Moon landing was completed. It was obvious that the space program wouldn't have the same high level of funding as Apollo, however, it was still important to look beyond the Moon and to future projects that could benefit America and preserve the country's leadership in space. Political forces and domestic needs would drive the NASA budget and the scope of the future projects in space exploration. Even though national support dwindled after the Moon landing, NASA still felt that Skylab would provide a way of maintaining a presence in manned spaceflight while learning about living and working in space. The next generation of hardware and mission objectives could be formulated while Skylab was performing its experiments. But congress wasn't interested in funding Skylab.[16]

The air force came to the rescue for NASA. It was already working on the Manned Orbiting Laboratory (MOL) that would use the upper stage from a Titan II rocket and a modified Gemini space capsule as an orbiting spy station, carrying out long-term surveillance missions on the Soviet Union using cameras, infrared sensors, and radar. The NASA Skylab plans offered a much larger platform with more sophisticated equipment, so it was thought that the MOL and Skylab could be combined, and Skylab was given the go-ahead.

By 1971, though, the Air Force decided that its spy cameras and surveillance equipment would be more flexible if deployed on unmanned satellites that could operate permanently in polar orbits. The MOL program was canceled, and most of its "spy" astronauts were folded into NASA. Skylab continued without military funding and was launched in May 1973. There were some issues with the solar panels at the start, but once that was addressed, Skylab continued operating successfully for several years before it was deliberately de-orbited with a re-entry burn over the Australian outback (Fig. 7.2).[17]

With the success of the Skylab missions, NASA thought that Skylab could be turned into a much larger permanently manned space station, which could be resupplied by the newly planned space shuttle. The proposed space station could serve as a steppingstone to explore Mars. Once again, politics affected

[16] Compton, W. David and Benson, Charles D. The NASA History Series. [Internet] history.nasa.gov; SP-4208 Living and working in space: a history of Skylab; 1983 [cited 2015 Sep 24]. Available from: http://articles.adsabs.harvard.edu/full/seri/NASSP/4208//0000001,004.html

[17] Flank, Lenny. [Internet] Dailykos.com; The sky is falling: the life and death of Skylab; Apr 16, 2014 [cited 2015 Sep 24]. Available from: http://www.dailykos.com/story/2014/04/16/1250880/-The-Sky-is-Falling-The-Life-and-Death-of-Skylab

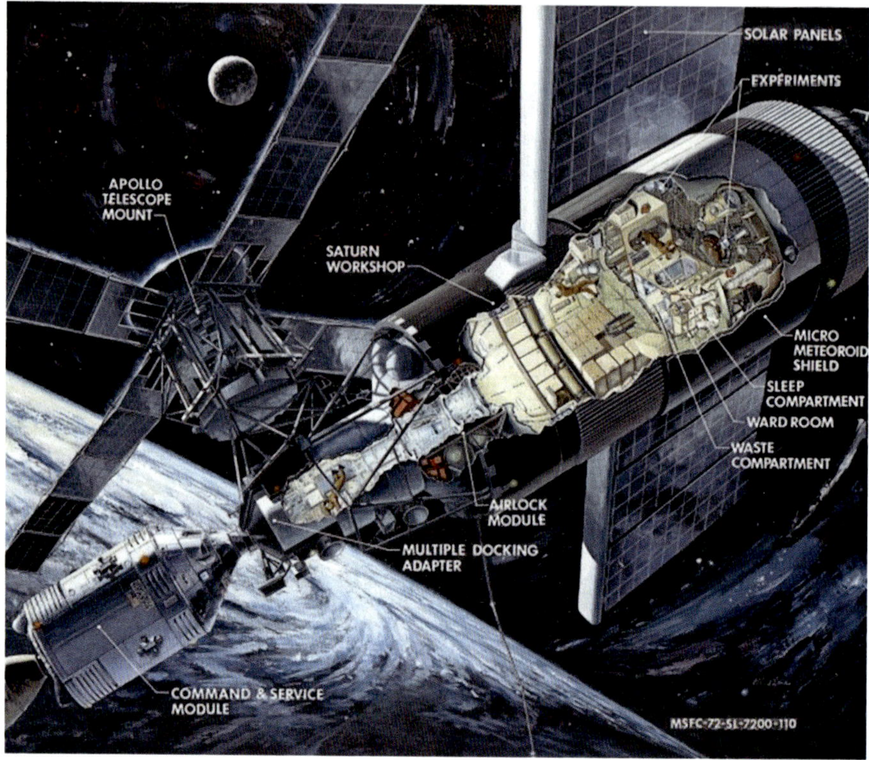

Fig. 7.2 Skylab cutaway (Image courtesy of Marshall Space Flight Center, NASA)

this plan. In 1972, President Nixon announced that the space shuttle would be funded, but the space station would not (leaving the space shuttle without a clear mission, a political handicap that would affect its future for the next four decades). It was intended that the first shuttle missions, planned for the mid 70s, would be used to resupply Skylab, and install an external booster rocket to push the station to a higher orbit, giving the station another 5 years of life. However, the space shuttle was delayed, the booster was never installed, and Skylab fell back to Earth. Skylab became another example of a time when progress was halted due to short-sighted thinking.[18]

[18] Flank, Lenny. [Internet] Dailykos.com; The sky is falling: the life and death of Skylab; Apr 16, 2014 [cited 2015 Sep 24]. Available from: http://www.dailykos.com/story/2014/04/16/1250880/-The-Sky-is-Falling-The-Life-and-Death-of-Skylab

The Space Shuttle

After the Moon landing, NASA's focus in the late 1960s was to keep America as a leader in space. A reusable space shuttle vehicle was proposed as a low-cost manned transportation vehicle that would operate between the surface of Earth and low Earth orbit, docking with a manned space station (an updated, bigger, and more permanent version of Skylab). The space station's primary goal was to be able to provide the next step into outer space. As stated previously, the ultimate decision to go forward with the shuttle program followed a series of changing priorities made by congress and President Nixon.

It became obvious during 1970 that NASA would not be funded for both the shuttle and a space station. NASA thought that the space shuttle would have a better chance of being funding over a space station, so the focus during the 70s was the development of the shuttle program. NASA was asked to justify the shuttle development by identifying its future roles, applications, and users, including the Department of Defense, commercial users, and other countries. The emphasis was on a financial justification rather than technical reasons for space exploration in the long term and how to accomplish those goals. The future of space exploration was in jeopardy and restricted by available funding.

The allotted budget in 1971 was set and restricted for the duration of the program ($3.2 billion for the duration despite the high required operational costs). Design was now tied to cost estimates, and some radical changes had to be made to afford the program. Finally, in early 1972, President Nixon and government advisers approved the development of the space shuttle as a means to remain a leader in space for both manned and unmanned missions throughout the 1980s and beyond. The budget was restricted, limiting the shuttle's capabilities. The original goal of having a vehicle system that was 100% re-usable was never achieved. (Although the solid rocket boosters were refurbished and reused, the external fuel tank was jettisoned into the ocean and not recovered.) NASA's space exploration goals and government funding became tied together. Each line item became a bargaining chip for political favors and compromise. The result was a space shuttle program that has been criticized for its lack of vision. NASA's original priority was to build one or more large manned space stations that would conduct scientific research focused down toward Earth and out into outer space. The International Space Station was also envisioned to be the steppingstone to other planets. The space shuttle was to be created as the transport system to support the ISS. Without the space station existing until the end of the shuttle era, the vision was out of sequence. When the ISS was finally built, the shuttle program ended, and the United States was left to depend on Soviet spacecraft to transport crews to and from the ISS.[19]

The shuttle operated for 30 years, from 1981 until 2011. The orbiter design was first envisioned by Max Faget whose life and accomplishments are discussed next.

The Visions of Max Faget

Maxime A. Faget (1921–2004) was an aerospace research scientist working for NASA Langley as far back as 1946. He was a designer and a visionary, transforming aerodynamic theory into practical aircraft and spacecraft designs. He was a member of the Pilotless Aircraft Research Division, focusing on high speed flight. In 1954, he took part in an initial feasibility study that led to the design of the X-15, the first hypersonic aircraft. He designed and proposed the original one-man capsule used in Project Mercury and contributed to the designs of every US manned spacecraft up through the space shuttle.[20]

"Without Max Faget's innovative designs and thoughtful approach to problem solving, America's space program would have had trouble getting off the ground," said NASA administrator Sean O'Keefe reflecting on Dr. Faget's contributions.[21]

In 1958 he became part of the Space Task Group, the newly formulated NASA group focused on manned spaceflight that later became the Johnson Space Center in Houston, Texas. Faget led the team doing the initial designs and feasibility studies to fly to the Moon. Working with a longtime associate, Caldwell Johnson, Faget devised a blunt shape for Project Mercury. His Mercury capsule took shape as a cone, with its broad end forward and covered with a thick layer of material to provide thermal protection. He came to Houston as a founding member of the Manned Spacecraft Center (MSC), where he became Director of Research and Engineering. He also adapted his basic shape to provide capsules for Gemini and Apollo.[22]

"Max Faget was truly a legend of the manned space flight program," said Christopher C. Kraft, former Johnson Space Center director. "He was a true

[19] Logsdon, John M. May 1986. The decision to develop the Space Shuttle. Space Policy. 2:2:103–119.

[20] History.NASA.gov. [Internet] History.NASA.gov; Oct 15, 2004 [cited 2015 Oct 05]. Available from: http://history.nasa.gov/Apollo204/faget.html

[21] NASA.gov. [Internet] NASA.gov; Oct 10, 2004 [cited 2015 Oct 01]. Available from: http://www.nasa.gov/vision/space/features/faget_obit_prt.htm

[22] NASA.gov. [Internet] NASA.gov; Oct 10, 2004 [cited 2015 Oct 01]. Available from: http://www.nasa.gov/vision/space/features/faget_obit_prt.htm

icon of the space program. There is no one in space flight history in this or any other country who has had a larger impact on man's quest in space exploration. He was a colleague and a friend I regarded with the highest esteem. History will remember him as one of the really great scientists of the twentieth century."[23]

Faget led the original feasibility study for the space shuttle design. Shuttle design studies in the late 1960s were focused on a Lockheed space place design developed by Maxwell Hunter called the Star Clipper. Faget worked on his own configuration at the Manned Spacecraft Center. His turned out to be a methodical approach that changed the future of the space shuttle program. Other design participants were focusing on whether the vehicle system would be fully or partially reusable rather than focusing on a specific design. Faget started with the concept of a two-stage fully-reusable shuttle that eventually gained popularity even in the debate over its specifics. As the debate continued, Faget was able to fine tune and optimize his design, shutting out competitors. The design was not revolutionary. He admits "My history has always been to take the most conservative approach." He was not a fan of the popular lifting body designs that could lead to aerodynamic interference between the wing and the body. However, a practical shuttle design would have to be a compromise between Faget's ideas and the lifting body capabilities. The aerodynamic challenges of the lifting bodies were that they had high drag and low lift and would come in to land at a very high speed. The body itself did the work of a wing. A small wing addition to the craft helped somewhat but was really a quick fix to a bigger problem.[24]

Lifting bodies were tested extensively by NASA with some success but had some very unstable flying characteristics. Some aspects would fit well with the space shuttle configuration, being able to fly like an airplane with sufficient lift to land as a glider. Re-entry into the atmosphere would require a blunt body nose with high drag and thermal protection underneath the body and wings. "With extremely high drag," he says, "you throw a big shock wave in front of you, and all the energy goes into that shock." The airplane would experience drag through friction with the atmosphere that would transfer heat to its surface. His design was fine-tuned into the shuttle design that was accepted and developed by NASA (Fig. 7.3).[25]

[23] NASA.gov. [Internet] NASA.gov; Oct 10, 2004 [cited 2015 Oct 01]. Available from: http://www.nasa.gov/vision/space/features/faget_obit_prt.htm

[24] History.NASA.gov. SP-4221 The Space Shuttle decision [Internet] History.NASA.gov; [cited 2015 Oct 05]. Available from: http://history.nasa.gov/SP-4221/ch5.htm

[25] History.NASA.gov. SP-4221 The Space Shuttle decision [Internet] History.NASA.gov; [cited 2015 Oct 05]. Available from: http://history.nasa.gov/SP-4221/ch5.htm

Fig. 7.3 Max Faget's space shuttle concept design (Image courtesy of NASA)

Max Faget retired from NASA in 1981, after the second shuttle mission, and founded one of earliest private space companies, Space Industries Inc. One of his projects, a small orbiting platform called an "industrial space facility," demonstrated the capability to process materials in a vacuum but was never utilized in commercial efforts. His contributions in spacecraft design, however, had contributed greatly to the success of the manned space program.[26] He was a visionary with the technical expertise to apply those visions to specific goals with a practical approach.

The Accomplishments of the Space Shuttle System and the Future in Space Transportation

There were six orbiter vehicles built, one flown only in Earth's atmosphere (Enterprise). Columbia flew the first five shuttle Earth-orbiting missions with extended length missions as long as 16 days. In the first three and a half years, only 24 missions were flown, well below the initial estimate for turnaround.

---

[26] Fox, Margalit P. 2004. Maxime Faget, 83; Pioneering aerospace engineer designed Mercury capsule. The New York Times; 8.

In 30 years, there were 119 shuttle launches, two ending in the loss of the crew and the destruction of the vehicles.[27]

Since the shuttle era ended in 2011, there have been mixed reactions to its legacy as well as a controversial decision to end the program prior to developing another vehicle able to transport astronauts to the International Space Station (ISS), leaving only Russian vehicles as transporters. It is important to reflect on what the space shuttle accomplished and what challenges are left for the next transport and delivery vehicle.

The primary accomplishment for the shuttle system was to prove a design to launch at least a partially reusable vehicle with the ability to enter orbit, perform routine space tasks such as delivering and repairing satellites, and have the capability to rendezvous with the ISS in order to deliver astronauts and supplies. All of these tasks were complex feats and accomplished successfully throughout the shuttle's lifespan. The shuttle never achieved the vision of a low-cost space "truck" with the original estimate of a quick turnaround of a couple of weeks between missions. However, the orbiter Discovery was launched nine times in 1 year,[28] and the shuttle system became the most cost-effective transportation system for launching payloads into Earth orbit. The space shuttle built the ISS, one piece at a time over a span of several years, starting in 1998. The ISS is discussed later in this chapter.

Some of the space shuttles most significant accomplishments during its 30 year history include:

- Three decades and 135 flights.
- 2300 experiments flown aboard to be tested in microgravity.
- Over 3.5 million pounds of cargo were launched into orbit.
- Almost 200,000 man-hours spent in space.
- Over 1300 days in space.
- 355 individual astronauts and cosmonauts flown, hailing from 16 different countries.
- 180 satellites and other payloads deployed (including components of the ISS) (Fig. 7.4).[29]

---

[27] NASA.gov. [Internet] NASA.gov; Oct 10, 2004 [cited 2015 Nov 11]. Available from: http://www.nasa.gov/sites/default/files/files/Spacelab_Collection_140117a.pdf

[28] Wall, Mike. [Internet]. Space.com; c2012. Space Shuttle Discovery: 5 surprising facts about NASA's oldest orbiter; April 19, 2012 [cited 2015 Nov 16]. Available from: http://www.space.com/15330-space-shuttle-discovery-5-surprising-facts.html

[29] CBS.news. [Internet]cbsnews.com; Space Shuttle: 30 years of fascinating facts. July 21, 2011; [cited 2015 Nov 15]. Available from: http://www.cbsnews.com/news/space-shuttle-30-years-of-fascinating-facts/

Fig. 7.4 STS-1, the first space shuttle launch (April 12, 1981) (Image courtesy of NASA)

Although the space shuttle's mission was never designated as being primarily scientific, it provided a laboratory in space to perform scientific research in a number of disciplines, including microgravity. Further research could then be performed on a much larger scale on the ISS. The importance of experimentation without gravity is similar to experimentation without air molecules (a vacuum). Experiments are created in vacuums on Earth to study processes without Earth's atmosphere. Research using vacuums and manipulation of air pressure led to the development of the light bulb, integrated circuits, freeze-dried foods, electron microscopes, particle accelerators, weather forecasting, and knowledge of human flight in air and space.[30] Research in a vacuum on Earth doesn't completely simulate the space environment. As discussed previously, the ISS and the space shuttle experience nearly zero gravity due to their free fall in orbit around Earth. It is important to study the long-term effects of no gravity in order to overcome challenges associated with human body responses for extended missions. Additional studies into the effects of space radiation on DNA and cells contributed to our knowledge of the long-term effects of humans traveling in outer space for extended periods of time. The space shuttle opened up studying disciplines previously limited in space. Research into biology and materials science expanded the knowledge of cell and crystal growth that advanced medical science and technology.

Microgravity research was conducted on Skylab, continued on the space shuttle missions, and expanded on the ISS along with other experimentation. ISS research will be detailed in a later chapter.

Space shuttle research was conducted primarily on the Spacelab module or other Spacelab experimental units placed in the orbiter payload bay.

The Spacelab Module

> "Shakespeare once wrote: 'Thoughts are but dreams 'til their effects be tried.' With Spacelab we have transformed the thoughts and dreams of thousands into reality."
>
> –James M. Beggs, NASA administrator, at the ceremony for Spacelab's arrival from Europe at the Kennedy Space Center in Florida, February 5, 1982.[31]

[30] Witze, Alexandra, Kenneth. 18 June 2011. Good-bye Shuttle: looking back at the space plane's scientific legacy. Science News Vol. 179. No. 13, pp. 20–21.

[31] Walter Froehlich. The NASA History Series. [Internet] history.nasa.gov; EP-165 Spacelab: Chapter seven: Spacelab: its birth, its impact, its future living and working in space: a history of Skylab; 1983 [cited 2016 Feb 22]. Available from: http://history.nasa.gov/EP-165/ch7.htm

In the late 1960s, Dr. Thomas Paine, the NASA administrator at the time, traveled to 19 countries to assess levels of interest in cooperating both scientifically and financially with the United States in space endeavors. NASA was looking at possibilities for post-Apollo programs, but all of the programs seriously considered were very expensive. One way to accomplish these projects was to make them worldwide cooperative efforts with international sharing of technology and funding.[32]

The Spacelab concept originated with the Space Task Group that was commissioned by President Nixon and chaired by Vice President Agnew. The group was directed to investigate the best ways of carrying out scientific objectives in space in the coming decades. Of the options presented, the space shuttle was chosen by President Nixon as the project he wanted to focus on. In addition, the Space Task Group made a strong recommendation to internationalize the space program.[33]

NASA's disappointment over its first bid for a space station in the early 1970s gave birth to the alternative concept of the Spacelab module. NASA immediately modified the Research and Applications Modules (RAM), which would enter orbit in the payload bay of the space shuttle, perform their functions as a stand-alone laboratory module, and return to Earth at the end of the shuttle mission.

Robert Lohman, NASA's chief of Spacelab development, said: "Once that decision was made in 1972, a lot of us were appalled that there was nothing left in the plan for space science. So we took the idea of these RAM'S [Research and Applications Modules, intended to be carried up and attached to the space station by the shuttle], and started to look at using them in the shuttle instead of for 'sortie' missions."[34]

NASA offered European countries the opportunity to partner in a Spacelab venture, who were enthusiastic about this idea. Discussions began between NASA and the two European space agencies of that time—the European Launcher Development Organization (ELDO) and the European Space Research Organization (ESRO). Both organizations were merged in 1975 into the European Space Agency (ESA).

---

[32] Space.com Staff. [Internet]; space.com; c2012. Timeline: 50 years of spaceflight. September 28 2012; [cited Feb 22 2016]. Available from: http://www.space.com/4422-timeline-50-years-spaceflight.html

[33] Walter Froehlich. The NASA History Series. [Internet] history.nasa.gov; EP-165 Spacelab: Chapter 7: Spacelab: its birth, its impact, its future living and working in space: a history of Skylab; 1983 [cited 2016 Feb 22]. Available from: http://history.nasa.gov/EP-165/ch7.htm

[34] Waldrop, M. Mitchell. AAAS Science Archives 1983–1985. [Internet] Spacelab: science on the shuttle. [cited 2016 Feb 22]. Available from: http://www.ganino.com/games/Science/Science%201983-1985/root/data/Science_1983-1985/pdf/1983_v222_n4622/p4622_0405.pdf

Discussions continued over several years and ended in an agreement for the European nations to develop a unique module/space laboratory that would utilize the shuttle's capacity to carry out scientific research. Both the technology and funding required were within ESA's means. The international agreement was signed by ten European partners in August 1973 (nine partners initially with Austria signing later). In addition, details about the overall Spacelab program operations were established between all partners. This was called the Spacelab Memorandum of Understanding (MOU). This agreement represented the first international technical and scientific cooperative agreement of this magnitude. It gave Europe the right to fund, design, build, and deliver Spacelab in exchange for a shared first mission aboard the space shuttle. In June 1974, the European Space Agency (ESA) selected an industrial consortium to develop the modular pieces to fit inside Spacelab, including a pressurized laboratory. The laboratory would provide the opportunities for businesses and universities to conduct a variety of research activities. Congress was already encouraging NASA to branch out into privatization and international partnerships to achieve common goals.[35]

The Spacelab project required detailed task delineation between all parties and complex management procedures. The first Spacelab flight was planned to be a cooperative mission, with NASA and ESA both flying experiments of equivalent magnitudes. There would also be a European scientist onboard as a crew member. Politics played a big role in establishing the conditions and criteria for the international team agreements and responsibilities.[36]

Unlike Skylab, the first US space station, which had been built mostly from existing Apollo hardware, Spacelab was a new construction offering a much wider range of applications. Spacelab was designed to fit into the shuttle cargo bay and connect with the crew compartment, allowing scientists to work in a pressurized laboratory in a shirt-sleeve environment. In addition, unpressurized external pallets would provide research platforms for external data collection and research in fields of astronomy, studies of Earth's atmosphere, and other observations. The lab was adaptable and reusable and was used in 26 missions over 16 years (1983–1998), conducting hundreds of experiments in the microgravity environment of low-Earth orbit. Space research took an

[35] NASA.gov. [Internet] NASA.gov; 2013 [cited 2015 Dec 29]. Available from: http://www.nasa.gov/sites/default/files/files/Spacelab

[36] Walter Froehlich. The NASA History Series. [Internet] history.nasa.gov; EP-165 Spacelab: Chapter 7: Spacelab: its birth, its impact, its future living and working in space: a history of Skylab; 1983 [cited 2016 Feb 22]. Available from: http://history.nasa.gov/EP-165/ch7.htm

important step into "hands-on" experimentation previously limited aboard Skylab or remotely via rockets or satellites controlled from Earth.[37]

The first flight of Spacelab took place aboard the space shuttle *Columbia* in November 1983. It was the first time a citizen of another country flew as an astronaut on the US spacecraft.[38] Spacelab turned out to be one of the most important and most frequently flown shuttle payload systems.

Spacelab became an intermediate step in the development of the ISS, allowing NASA to achieve several scientific objectives with the financial backing of ESA. By 1972, NASA had already postponed the development of a large space station due to the inability to fund both the space shuttle and a space station. Spacelab provided the opportunity to conduct space experiments in the interim.

NASA, specifically the Marshall Space Flight Center in Huntsville, Alabama, was responsible for the overall program planning and management of Spacelab while ESA designed and developed the module and pallets. Marshall was the logical choice to manage the project, having previously conducted a study to design a module suitable for short-duration flights and capable of Earth observation astronomy. Marshall became experienced in international space partnerships and missions and looked forward to planning similar modules for an International Space Station.[39]

Research accomplished in Spacelab through the shuttle resulted in major discoveries in astronomy, biology, and crystallography. These experiments paved the way for more in-depth experimentation aboard the ISS, discussed later.

Legacy of the Space Shuttle

Over the years, the space shuttle demonstrated its capabilities over the years to serve as a launch, delivery, and recovery system. It launched, recovered, and refurbished satellites, delivered space station modules, and transported supplies and crews to the ISS. In addition, the shuttle served as a platform for complex spacewalks and robotics work, observing both Earth and the universe with cutting-edge scientific payloads.

---

[37] NASA.gov. [Internet] NASA.gov; 2013 [cited 2015 Dec 29]. Available from: http://www.nasa.gov/sites/default/files/files/Spacelab_Collection_140117a.pdf

[38] Wilford, John Noble. 29 Nov 1983. Columbia carries spacelab to orbit with 6-man crew. The New York Times. [Internet] [cited 2016 Feb 23]. Available from: http://www.nytimes.com/1983/11/29/us/columbia-carries-spacelab-to-orbit-with-6-man-crew.html?pagewanted=all

[39] NASA.gov. [Internet] NASA.gov; 2013 [cited 2015 Dec 29]. Available from: http://www.nasa.gov/sites/default/files/files/Spacelab_Collection_140117a.pdf

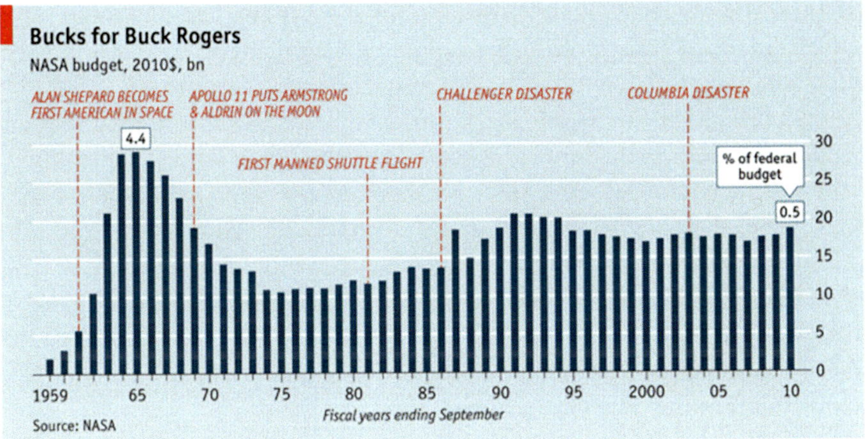

Fig. 7.5 NASA budget 1959–2010 (Image courtesy of NASA)

The shuttle provided sufficient lifting power with additional flexibility and versatility. It was an accessible classroom in space, teaching us that living and working in space could be a reality. The orbiter fleet demonstrated the vital skills of orbital repair, outpost construction, precision rendezvous and docking, complex EVA, and intense, round-the-clock scientific operations. Its technical problems and challenges will help build safer and more efficient vehicles in the future.

One of the most important achievements of the space shuttle was its role in launching and repairing the Hubble Space Telescope, the most sophisticated instrumentation yet developed to learn about the universe. The technical difficulties that occurred early on with the Hubble required the shuttle to perform a challenging technical repair in space. The successful in-orbit repair made it possible for the space telescope to capture incredible images of the universe for many more years, unfolding secrets of outer space never before seen from outside of Earth's atmosphere.

Estimates for the cost of the shuttle program and individual launches vary, but NASA claims that each launch cost $450 million, with others saying that it could be more like $1–$1.5 billion. This is estimated to be much more expensive than Russia's Proton rockets that provided a comparable transport system but at about a quarter of the cost because the rockets were older (60s) technology and expendable.[40]

NASA's budget for the past 50 years is shown in Fig. 7.5. The biggest spike in terms of percent of the federal budget understandably occurred during the Cold War (4.4 % at the peak). The budget dropped dramatically after that and

[40] The Economist (London, England), The Space Shuttle, into the sunset, July 02, 2011; p. 9; Issue 8740.

flattened out at a level much lower level for the last 20+ years (close to 0.5 % of the budget). With priorities being placed elsewhere, there is not a lot of federal monies available today for NASA and outer space travel.

One criticism of the shuttle program was that it was short sighted, because it didn't plan for manned missions outside of low Earth orbit.

"It is now commonly accepted that was not the right path," then-NASA chief Michael Griffin told *USA Today* in 2005, referring to the low orbit activities. "We are now trying to change the path while doing as little damage as we can."[41] However, the new path that Griffin referred to was President George W. Bush's Moon mission called Constellation, which was subsequently canceled by President Obama as being too costly during a time of Middle Eastern wartime expenditures.

The space shuttle was a great national accomplishment for the United States, furthering scientific research in microgravity and the effects of extended periods in space on the human body. The shuttle was, without dispute, an important part of American space exploration goals. However, rather than moving along with consistent goals and development of propulsion and travel vehicles, the space program became a line item under the discretion of short-sighted politicians and partisanship. This political partnership became both a benefit in the space race and a curse later on, when specific goals and planning were required to move into the next phase of space exploration, going to Mars. Now, decades after landing on the Moon, preparations for a trip to Mars are requiring all new technology and transport systems.

Once again, it is obvious that any organization must have consistent planning and funding to achieve its goals over time, and NASA, shepherding the future of the space program, is no exception. The space program requires more than federal budgeting; it requires a groundswell of interest in space exploration, lending support for increased spending for space exploration as well as inspiring more private enterprise to explore outer space.

The Military Influence on Space Shuttle Operations

The US Department of Defense (DOD) took a strong interest in the mission applications for the space shuttle. The air force was tasked to work with NASA in developing the shuttle system. The cargo bay of the shuttle was designed to hold spy satellites, and up until the Challenger disaster, the military was

[41] Wall, Mike. [Internet]. Space.com; c2011. NASA's Shuttle program cost $209 billion—was it worth it?; July 05, 2011 [cited 2015 Nov 11]. Available from: http://www.space.com/12166-space-shuttle-program-cost-promises-209-billion.html

using the shuttle's payload bay to transport surveillance equipment. Although NASA promoted the space shuttle as a civilian vehicle, DOD agreed to support and partner with NASA as a means for military operations in space. The immediate need was to deploy and use reconnaissance and national security payloads in low-Earth orbits. Design modifications were required to support the military space program, otherwise the DOD could have withheld political and financial support for the project. To gain this support, NASA focused on accommodating military missions. Having DOD support was critical to overall government support and funding. The shuttle was pitched to President Nixon as an essential part of national security.

Since the Cold War and the space race, outer space was considered to be a competitive environment for military operations, including surveillance and communications. A space transportation system fit well into low orbit satellite operations while providing a platform for prototype systems, perhaps even weapons. For the DOD, the space shuttle was a win-win. The shuttle would provide cheaper and more flexible options for military space operations. The deployment, repair, and retrieval of satellites were attractive capabilities to achieve efficient performance in communication and navigation systems.

Proposed military missions would require deployment of satellites on occasion to high inclination orbits for surveillance of certain regions on Earth. Military launch facilities at California's Vandenberg Air Force Base would be overhauled and fitted for these missions. The base location was perfect for launching shuttles over the ocean to reach polar orbit, the destination for surveillance and imaging satellites. Reaching that type of orbit would not be possible from Florida's Cape Canaveral because it would require the shuttle to fly over populated land after launch. The US Air Force space shuttle era was supposed to begin in 1986 with astronaut and commander Bob Crippen (first crew of the space shuttle) to be commander of the maiden mission to polar orbit carrying the Teal Ruby experimental satellite along with long range sensors in the payload bay. Expectations for the shuttle to be an integral part of military space operations were high. To prepare for this, between 1979 and 1986, DOD trained 32 navy and air force officers as military astronauts. In 1986, the DOD started a Military Man-in-Space Program to make sure that a human military presence remained in space. It was believed that an experienced military astronaut's judgment would be necessary when dealing with complex situations.[42]

Soon after the shuttle's first launch, it became clear that there would be problems with easy turnarounds and multiple launches per year. The military

---

[42] Ray, Justin. [Internet]. Space.com; c2011. From Shuttles to rockets: long history for California launch pad; January 19, 2011 [cited 2015 Dec 30]. Available from:http://www.space.com/10644-california-launch-pad-history-shuttles-rockets.html

looked more closely at their plan to use the shuttle for military operations. The cost advantage was re-evaluated over expendable launch vehicles, and it no longer seemed viable to depend on the space shuttle to transport military systems into space. Unmanned booster operations were continued until it was shown that the shuttle could meet the demands of the military. A top ranking military official (Under Secretary of the Air Force Edward "Pete" Aldridge) was chosen to fly aboard the first Vandenberg shuttle mission. After the Challenger explosion, the Vandenberg shuttle missions were canceled, and the Pentagon focused on developing expendable rockets for their payload needs. Only payloads requiring astronaut assistance would fly aboard the shuttle.[43]

The DOD began work along with NASA in the 1980s on a single-stage-to-orbit (SSTO) vehicle for military purposes. It appeared that the space shuttle was not able to deliver on its expectations, and so the DOD proposed the development of a hypersonic space plane that could take off, fly into orbit, perform its mission, and return like an airplane. A proposal was submitted to the Defense Advanced Research Projects Agency (DARPA) and funded as a secret program between 1983 and 1985. The Reagan administration announced it as the National Aerospace Plane, designated as the X-30. The design was sophisticated and challenging. After billions were spent, the project ended in 1994 amid scoreless technical difficulties. However, the concepts of this program and this type of vehicle remain today as an important component in military space defense and aggressive war capabilities. A military presence in space was still considered to be part of a strategy essential for national security.[44]

There were a few shuttle missions that were classified. Between 1982 and 1992, NASA launched 11 classified payloads, utilizing changes requested by the military of the cargo bay. Of all of the military astronauts, only one made it to orbit, Gary Payton, who became the deputy undersecretary of the air force for space. He recalls the tension that existed between NASA and the military. The military astronauts were payload specialists, engineers, or scientists who focused on a particular experiment or satellite and typically flew only once. They didn't bridge the gap between NASA and the military. Despite differences, military payloads were flown and launched successfully and individual mission requirements were satisfied.[45]

---

[43] Ray, Justin. [Internet]. Space.com; c2011. From Shuttles to rockets: long history for California launch pad; January 19, 2011 [cited 2015 Dec 30]. Available from:http://www.space.com/10644-california-launch-pad-history-shuttles-rockets.html

[44] Launius, Roger. [Internet]. wordpress.com; c2012. NASA's Space Shuttle and the department of defense; Nov 12, 2012 [cited 2015 Dec 30]. Available from: https://launiusr.wordpress.com/2012/11/12/nasas-space-shuttle-and-the-department-of-defense/

[45] Cassutt, Michael. Air & Space Magazine, Secret Space Shuttles, August, 2009.

8

Politics, the ISS, and Private Enterprise

Keywords Boeing CST-100 • Charles Bolden • President Clinton • International Space Station (ISS) • Orbital Sciences • Orion Multi-Purpose Crew Vehicle • President Reagan • Space Exploration Initiative • Space Shuttle • Space Station Freedom • Space Task Group • SpaceX Dragon V2

> "During the next 50 years, in countless cycles, in countless entrepreneurial companies, this 'let's just go and do it' mentality will help us final get off the planet and irreversibly open the space frontier. The capital and tools are finally being placed into the hands of those willing to risk, willing to fail, willing to follow the dreams."
>
> –Dr. Peter H. Diamandis, chairman of the X-Prize Foundation, "The Next 50 Years in Space," *Aviation Week* online, March 14, 2007

As we discussed earlier, the space shuttle program demonstrated the ability to transport payloads and humans into space and to land safely back on Earth. The public was initially very excited about the new space technology used to develop a multi-use manned reusable spacecraft. All of the shuttle's activities and research occurred in low-Earth orbit. But once again, the initial excitement of humans returning to space started to dwindle in the eyes of the public after several successful missions. Inevitably, routine activities conducted in space fell into the background of American daily life. Often, the only time that the public took notice was when a crisis occurred in the program. Despite this, NASA had enough foresight to plan and finance the development of the International Space Station (ISS). With the help from politicians acting as

L. Dawson, *The Politics and Perils of Space Exploration*,
Springer Praxis Books, DOI 10.1007/978-3-319-38813-7_8

salesmen for NASA, the United States was able to acquire international partners that made the design and development of the station possible.

This chapter will examine the politics of the US space program both nationally and internationally in the years following the space shuttle era. The development of the ISS is addressed as well as private enterprise's emergence as a vital component in the future of American space exploration.

The ISS's role in continuing the success of the space program is known and appreciated by scientists and others who understand the research necessary to learn about human physiology in microgravity and the challenges of long-term space missions. The public is generally not well informed about scientific discoveries associated with space exploration. Only a small percentage of the population even realize that there are astronauts and cosmonauts continually working and living in the space, much less appreciate the importance of the research being conducted on a continual basis.

Why is the lack of knowledge and lack of appreciation of these efforts on the public's part important? Let's remember our premise that public groundswell support for publicly funded efforts could provide an important rationale for the government's prioritization of funding. In times of financial stagnation or economic depression, some efforts can and do suffer a loss of support. But, if powerful interest groups for one research area, such as climate change, are loud enough, they can and will influence government sources of funding through strong debate and public show of support.

Increased support during the space shuttle era was met by private industry and entrepreneurial activities that focused on missions for transport to and exploitation of low-Earth orbit and beyond. We are now poised and prepared to make the next big leap in space travel.

International Space Station Politics

The ISS is arguably one of the greatest technological achievements of humankind. It demonstrated what could be accomplished with international cooperation. Led by the United States, the ISS program began in 1982, with assembly beginning in 1998, requiring over 30 space missions and about 15 years to complete. It is a shining example of how different space agencies can plan and coordinate the operations of a very complex system.

The history of the priorities set by the Space Task Group and the resulting decisions made by President Nixon were outlined earlier. The recommendations from this group were to continue to explore space for peaceful purposes and to include manned space missions as an important component of the pursuit of certain scientific objectives, including gaining a better understanding of the universe. Another objective was the development of a low-cost, reusable space transportation system that would study Earth and space and supply a modular space station created with international cooperation and interests.[1] President Nixon chose to develop the space shuttle and remain in low-Earth orbit, abandoning the other recommendations as too expensive and far-reaching.[2]

Once the space shuttle was developed and operational in 1982, NASA started a conceptual design for a large manned space station, built over time from individual components. It would utilize the capabilities of the space shuttle and serve as an intermediate base for further exploration of the Moon and other planets. Emphasis was to develop the project with international cooperation, both technical and financial. And so, NASA invited Canada and US-friendly European countries to participate in the project. In addition, NASA's director Beggs invited Japan to participate. Canada had already partnered with NASA's shuttle program to develop the remote manipulator arm mounted in the payload bay of the orbiter to perform tasks such as deploying and rescuing satellites. In addition, ESA, a consortium of European countries dedicated to space research and developing space systems, was also interested in a long-term project for space observations and experiments. They had already participated in the development of Spacelab, the manned space laboratory module that fit into the shuttle cargo bay and performed a number of experiments on several shuttle flights.

President Reagan declared his support for the construction of a permanently manned Earth orbiting space station in his State of the Union address in 1984:

> America has always been greatest when we dared to be great. We can reach for greatness again. We can follow our dreams to distant stars, living and working in space for peaceful, economic and scientific gain. Tonight, I am directing NASA to develop a permanently manned space station and to do it within a decade.[3]

The manned space station was promoted as a technological achievement that could strengthen the economy, perform cutting-edge scientific research, and improve the quality of life. By 1985, Japan, the European space program,

[1] Space Task Group (US). The Post-Apollo space program: directions for the future. [Internet]. History Office, NASA Headquarters, Washington, DC; 1969 [cited 2015 Sep 21]. Available from: http://www.hq.nasa.gov/office/pao/History/taskgrp.html

[2] Logsdon, John M. After Apollo? Richard Nixon and the American space program. NY: Palgrave Macmillan; 2015. 368 p.

[3] Scimemi, Sam. NASA and the legacy of the International Space Station. NASA Advisory Council HEO Committee; July 29, 2013.

and Canada had decided to participate. The ISS went through a design process that included the concept of using it as an intermediate base for the exploration of other space and as a repair facility for satellites. The space station was named Freedom and, although initially approved, was never completed as originally designed and went through several cutbacks before evolving into the current International Space Station.

In 1989, through the Space Exploration Initiative, President George H. W. Bush outlined the construction of the space station Freedom, plans to return humans back to the Moon, as well as future plans for manned missions to Mars. He framed the vision in a similar way to how the space race began in 1961. However, the Apollo successful in landing on the Moon had government and public support, due to the Russian progress in space and the added political tensions. The president said he was inspired to seize an opportunity rather than respond to a crisis. The new plans, estimated to cost approximately $500 billion, spread over 20–30 years, was opposed by the White House and Congress primarily due to the cost. President Bush sought out international partners, but the program proved to be too expensive even with international support.

After Congress rejected the expensive proposal in 1990, the president established the Advisory Committee on the Future of Space program to make recommendations about how NASA should proceed. The committee felt that NASA should focus on Earth and space science using robotic methods, essentially ending the development of any new manned missions indefinitely. President Bush ordered NASA to go ahead with these recommendations. Dan Goldin was brought in as the new administrator, following the new philosophy of no human exploration beyond Earth orbit, and low-cost methods were applied to robotic missions. In 1996, President Clinton's National Space Policy removed human exploration from the US national agenda.[4]

Although the far-reaching plans were scrapped, strong interest in a low Earth orbit manned space station remained strong. Several designs were studied before the final configuration was chosen and the future missions and experiments defined. In 1993, President Clinton's administration partnered with Russia to help with the ISS construction and transport of support items to the station.[5] This collaboration was essential for the project to move forward. The Russians were able to help both financially and technically. They supplied several mod-

---

[4] Dick, Steve. Summary of space exploration initiative. [cited 2016 Jan 18]. Available from: http://history.nasa.gov/seisummary.htm

[5] Smith, Marcia S. NASA's space station program: evolution and current status: Testimony before the house science committee; Apr 4, 2001 [cited 2016 Jan 18]. Available from: http://history.nasa.gov/isstestimony2001.pdf

ules as well as the spacecraft required to transport the crew and cargo into orbit. In addition, the Russian space program could progress without taking on the sole financial burden of building its own space station. Soyuz spacecraft would transport crews, but the space shuttle was key to delivering the modules, using the robotic arm to help connect the modules. A Russian rocket launched the first piece of the ISS, and 2 years later, in 2000, the first crew arrived. Humans have occupied the space station ever since that time.

The US relationship with Russia or, previously, the Soviet Union, had been tense through the Cold War and the space race. After the United States landed on the Moon, the Russians started to focus more on studying humans working and living in space. With tensions eased between the two countries, there was increased cooperation and interest in sharing at least a symbolic moment ending the space race. This was accomplished in 1975 with the docking of an Apollo vehicle with a Soyuz vehicle in low Earth orbit. Several years passed before joint activities between the nations would resume. In 1992, after the collapse of the Soviet Union, President George H. Bush renewed the cooperative space efforts with agreements to launch a cosmonaut on the space shuttle while sending an American astronaut to Russia's space station MIR. Within a couple of years, shared activities would also include training of US astronauts in Russia.[6]

When President Bill Clinton took office, he directed NASA to once again redesign the space station, replacing Freedom with a less expensive International Space Station. Millions had been spent on Freedom, with little to show for it. Congress voted to support the ISS by a narrow margin. Clinton arranged for Russia to participate as a partner, not just a supplier of parts, boosting both their commitment and their financial support to the effort. The number of joint missions between the nations expanded, resulting in a perceived merging of the two space programs.

It is thought that in addition to financial and scientific support, Clinton's purpose of including Russia as an essential player in the ISS construction also had its roots in foreign policy. At the time, there was concern about Russia's position on ballistic missile proliferation. Russia's agreement on the proliferation policy happened at almost the same time that an announcement was made that they would become partners in the ISS operation and construction.[7] There were opponents (US scientists, astronauts, and public and

[6] Smith, Marcia S. NASA's space station program: evolution and current status: Testimony before the house science committee; Apr 4, 2001 [cited 2016 Jan 18]. Available from: http://history.nasa.gov/isstestimony2001.pdf

[7] JAXA. Japan Aerospace Exploration Agency. May, 1999 [cited 2016 Jan 18]. Available from: http://iss.jaxa.jp/iss/history/index_e.html

industry leaders) to Russia's increased profile in the US space program. As it turned out, they were justified in having doubts when Russia had difficulties fulfilling their financial commitments. The Russian module was the essential critical module of the station. The Russian progress was so delayed that NASA had decided to build its own makeshift component that could be put into place and allow the station plans to go forward. The delayed Russian module was to provide life support for crews as well as to propulsion and control for the orbital complex, so technically, it was very important to the life of the ISS. Eventually all the issues were resolved.

Projects of this scope always seem to expand, and this one grew both in vision and cost. The ISS was an incredibly ambitious project, the structure as big as a football field with connecting modules providing a livable environment for astronauts. There were many issues that were unable to be predicted or estimated both in time and financial commitment, and cost overruns were inevitable. The station progressed in stages, years passed, and construction wasn't completed until 2011. Construction was delayed for two and a half years when the space shuttle was grounded after the shuttle *Columbia* disintegrated on re-entry early in 2003. Aside from this period, the construction continued to completion.

Having reached the original design milestone—15 years of continuous operation—the ISS faced another budget hurdle and decision whether to extend its operations into 2024. President Obama announced that the ISS would extend operations, and NASA would continue spending upwards of $4 billion per year to keep the station functioning. Financial support by the international partners would be uncertain at best. In addition, the viability of all critical systems would have to be evaluated and determined if and what upgrades would be required. Astronauts routinely conducted spacewalks to fix critical components. Issues with degrading solar arrays affected the ability to generate power and could be a problem with increased longevity. Most importantly, the retirement of the space shuttle limited the US ability to deliver supplies, larger replacement parts as well as transporting and replacing crew members. US astronauts were ferried by the Russian Soyuz spacecraft, a controversial decision for NASA. It would cost upwards of $70 million to transport an astronaut to the ISS on the Soyuz, which is a big expense when you are exchanging four astronauts. Political tensions increased between the two nations as Putin postured and intimidated nearby countries during his administration as Russia's prime minister. Eventually, the situation calmed down, and in early 2015, the two countries once again agreed to partner to build a new space station after the ISS finished its extended life in 2024. Other international partnerships would be solicited before that next milestone. In addition to the space station agreement, Russia and the United States started to plan for a joint mission to Mars. NASA's Chief Charles Boden confirmed the partnership: "Our area of cooperation will be Mars. We are discussing how best to use the resources, the finance, we are settling time frames and distributing efforts in order to avoid duplication." [8]

Politics continues to limit NASA's objectives and future planning in space. The most significant difference today is the emergence of private enterprise in space, and NASA contracting out the task of transporting crew and cargo to private companies. These efforts will bring out new faces and support for deep space exploration that is not as dependent on national funding.

The Legacy of the International Space Station

The first step in launching the ISS began in 1998 with the launch of a Russian Proton rocket from Kazakhstan with an ISS module called Dawn (Zarya), which was funded by the United States and built by Russia. Only two weeks later, the space shuttle *Endeavor* delivered the US-built Unity module into space, connecting it with Dawn. ISS construction had begun.

The international cooperation required to build the ISS cannot be underestimated. It was very difficult to have so many nations cooperate on such a large project and maintain the common goals. The ISS presented the perfect opportunity to combine resources to construct a low orbit long-term space station devoted primarily to peaceful scientific objectives. It became the most expensive manmade structure ever, with a cost estimated close to $150 billion.[9] The planning, building, and operation of the station was a logistical success, although it was not always an easy road to completion. Its construction demonstrated the ability to build and connect modules in the hazardous environment of outer space (Fig. 8.1).

The challenge of a project of this magnitude was multi-faceted. There were enormous technical and logistical difficulties but also national and international political tensions that could affect funding. Because of the enormity of these challenges, you might wonder why a project of this magnitude was pursued, and why a manned space station such as the ISS is so important. These reasons are nearly the same today as they were at the start. A low-Earth orbit space station is an accessible laboratory on the edge of space with research that can

[8] rt.com news. Russia & UW agree to build new space station after ISS, work on joint Mars project. 28 Mar 2015 [cited 2016 Jan 18]. Available from: https://www.rt.com/news/244797-russia-us-new-space-station/

[9] LaFleur, Claude. Costs of US piloted programs. The Space Review. 2010 March 8.

Fig. 8.1 The ISS after STS-124 in 2008 (Image courtesy of NASA)

improve life on Earth. There have been results of experiments in medical studies, physical science, and the development of new materials as well as providing an opportunity for the long-term study of humans in zero gravity. The ISS has proven its ability to deliver results in all these areas. As an intermediate testing ground for missions into deep space, the ISS provides an important easily accessible location for evaluating equipment for use in long-term manned or unmanned missions. Lastly, the ISS can be a destination for new spacecraft from other countries and private companies as previously mentioned.

Eventually, the ISS will be de-orbited safely after being in continuous orbit for almost three decades. Research on human physiology and living and working in space will give scientists a strong basis for preparing to explore deep space.

Post-International Space Station Politics

NASA lost its ability to transport astronauts into space after the space shuttle retired in July of 2011. The Orion Multi-Purpose Crew Vehicle will be the next NASA transport spacecraft, but it won't be ready for crewed missions until after 2021. In the interim, NASA's only option for ferrying crews is to pay for rides aboard the Russia's Soyuz capsule. Private firms expressed interest and began development to provide this capability, although it is estimated that it will be at least 2017 before these vehicles are ready. Both SpaceX and Orbital Sciences won contracts through NASA to transport cargo to the ISS and provide low-Earth orbit access for a lower cost. NASA could now focus on deep space exploration and more far-reaching objectives. In 2014 NASA administrator Charles Bolden announced that Boeing and SpaceX will build the first private vehicles (Boeing CST-100 and SpaceX Dragon V2) for the purpose of launching American astronauts to the ISS, restoring the capability to launch crews from American soil for the first time since 2011.[10]

Historically, most of NASA's budget was paid to private contractors, to design and build space vehicles, rockets, and other equipment. NASA provided oversight, management, and operations of the overall projects. Today, NASA has moved to privatize some of the operations that focused on transport and low-Earth activities. This shift is an important one, affecting the future capabilities of NASA and the focus of future deep space exploration. In some ways, the future is uncertain, and some experts debate whether private industry can handle the complexity and safety of manned spaceflight beyond low-Earth orbit. What is certain is that the future of space exploration will be supported by a number of different resources. The public and NASA has realized that future missions, both manned and unmanned, will need to access resources from private sources and international partners, in addition to the traditional government channels. Another certainty is the public interest in space activities as evidenced through an increase of science fiction writing and movies and the numbers of private industry companies devoted to both manned and unmanned exploration of space. It will be interesting to see how it plays out in the next couple of decades.

[10] Kremer, Ken. Boeing and SpaceX win NASA's 'space taxi' contracts for space station flights. 17 Sep 2014 [cited 2016 Feb 25]. Available from: http://www.universetoday.com/114247/boeing-and-spacex-win-nasas-space-taxi-contracts-for-space-station-flights/

9

Technological Risks and Accidents

Keywords Antares rocket built by Orbital Sciences • Apollo 1 • Apollo Command and Service Module • Burt Rutan • Challenger • Christa McAuliffe • Columbia • Columbia Accident Investigation Board (CAIB) • Ed White • Elon Musk • Falcon 9 • Gus Grissom • John Glenn • Michael Smith • Morton-Thiokol • O-rings • President John F. Kennedy • Roger Chaffee • Sir Richard Branson • SpaceX • Thermal Protection System (TPS) • Virgin Galactic's SpaceShipTwo • WhiteKnightTwo • Zipper effect

> "We fooled ourselves into thinking this thing wouldn't crash. When I was in astronaut training I asked, 'What is the likelihood of another accident?' The answer I got was: one in 10,000, with an asterisk. The asterisk meant, 'we don't know.'"
>
> –Bryan O'Connor, Former Astronaut and Deputy Associate Administrator for the space shuttle, 1996

Every manned space disaster is a tragic and expensive loss. We feel a connection with human explorers who leave this planet to travel into the unknown of space. When something goes terribly wrong, it affects all of us who want to solve the mysteries of space and perhaps discover a clue to our own existence.

Poignant and terrifying moments recorded during past tragedies expose the danger and risk of space travel—the horrific screams of the *Apollo 1* astronauts as they were consumed by fire and then silence, the final "uh oh" uttered by *Challenger* pilot Michael Smith just prior to the shuttle explosion, and the video of the *Columbia* crew performing normal activities prior to its disastrous breakup on re-entry. With horror and fascination, we watched these events and tried to process what had happened and what it meant to the

L. Dawson, *The Politics and Perils of Space Exploration*,
Springer Praxis Books, DOI 10.1007/978-3-319-38813-7_9

future of the space program. Each accident investigation involved design and decision regrets and the eventual assessment and redesign.

Spaceflight is difficult, risky, and very expensive. Human spaceflight requires multiple backup systems in addition to the ability to return home safely along with rescue or abort plans. America has lost 17 astronauts in the line of duty in spacecraft. Several more cosmonauts have been lost in the Russian space program. Risk is there every time a rocket or space plane is launched. More lives will be lost in the future as we move forward to Mars missions, private enterprise endeavors, and space tourism. Human exploratory endeavors are inherently risky on their own. Add to that, new technologies, explosive fuels, and a hostile environment, and it is amazing that so few lives have been lost.

In his retirement speech in 1997, John Glenn said "I guess the question I'm asked the most often is: 'When you were sitting in that capsule listening to the countdown, how did you feel?' Well, the answer to that one is easy. I felt exactly how you would feel if you were getting ready to launch and knew you were sitting on top of two million parts—all built by the lowest bidder on a government contract."[1] It is probably inevitable that human errors, either in manufacturing or decision making, will occur in some aspect of the design or build procedures. Putting crew safety first requires extra levels of scrutiny, rigorous testing, and oversight that can be costly and cause delays in flights. At some point, decisions are made on a high level that say 90 % or 95 % safe is good enough.

Unmanned mission failures are also significant to those who have invested in the scientific result of their efforts. In addition, the success of manned missions often depends on unmanned missions, such as scouting landing areas and supplying provisions. The result of unmanned mission failures can be program delays, financial setbacks, and risk to humans those missions are supporting. Recent failures include a SpaceX rocket explosion during launch, a vehicle that was bringing supplies and experiments to the ISS. An Antares rocket built by Orbital Sciences blew up shortly after lifting off from a NASA launch facility on the shores of Virginia. It was also carrying a vital payload to the ISS. Even the future of space tourism has suffered setbacks. Virgin Galactic's SpaceShipTwo crashed during a test flight over California, killing one of the two pilots aboard.

In this chapter, the perils of space travel are described in terms of technological challenges and risks. Understanding the risks involved is an important piece to telling the story of current and future space exploration.

---

[1] Historicwings.com Staff. [Internet] history.com. Columbia disaster; c2015 [cited 2015 Aug 23]. Available from: http://www.historicwings.com/features98/mercury/seven-left-bottom.html

Technological Risks of Spaceflight

Scientists and engineers design spacecraft and spacesuits using the most current materials and technology available. Unfortunately, by the time the craft is launched or the object is in use, the technology is already several years old. Computer systems, technology, and materials change rapidly over time. The space shuttle's General Purpose Computer (GPC) was originally designed in early 1972 as a state-of-the-art flight computer, similar to those used on the F-16 fighter.[2] When the shuttle was launched in the 1980s, the computer systems were already considered aging technology and were not upgraded to more advanced capabilities until 1990.[3] Because technology changes so rapidly, it is difficult for manned spaceflight to maintain the most sophisticated and robust systems. Every technological risk is evaluated by the government or company organization in charge of that system, so it is fair to say that all of these systems and decisions that result are subject to human error.

The most explosive systems on space vehicles are related to fuels and launch systems. Rocket propulsion has not changed very much since the birth of space exploration. Solid rocket boosters are the simplest and are considered safe enough to use by hobbyists and their model rockets. Solid fuel contains a mixture of fuel and oxidizer that is burned until completion, releasing hot gases from a nozzle producing thrust. The fuel is an inert solid until ignited. Liquid fuels are volatile not only once mixed and combusted but also in storage tanks. The fuel burns when combined with an oxidizer, a substance that releases oxygen. The systems are complex, but the flow of propellant can be controlled or stopped. Hazards exist in the storage and transport of these chemicals. Going forward, new propulsion systems involving hydrogen or nuclear-generated fuels will still have similar safety and control issues.

All of the fatal accidents in the US space program were the result of mechanical or technological failures, but the systems and mission plans were designed by humans.

- The *Apollo 1* capsule was consumed by fire after a spark ignited some material inside of the capsule leading to a fire that consumed the capsule and exploded in a 100 % oxygen atmosphere. This accident happened during a ground test prior to launch.

[2] Chien, Philip. Space shuttle technology. Compute!. 1991 Aug; 132: p. 92.

[3] NASA, [Internet]. Spaceflight.nasa.gov; c2002. General-purpose computers; [cited 2015 Aug 15]. Available from: http://spaceflight.nasa.gov/shuttle/reference/shutref/orbiter/avionics/dps/gpc.html

- The *Challenger* space shuttle explosion was caused by a solid rocket booster O-ring leak.
- The *Columbia* space shuttle accident upon re-entry was caused by a failure of the heat resistant tiles on the wing, due to being compromised by a piece of foam striking it during launch.

NASA's investigation into these accidents exposed a number of system failures, poor engineering, and a lack of judgment. It is worthwhile to review a summary of these fatal accidents in order to look at how and if risk evaluation has changed in the space program.

Apollo 1 (January 27, 1967)

Historical Context

Due to increased funding and publicity, NASA was on track to land on the Moon by the end of the 1960s as promised by President Kennedy. During this time, however, tensions between the United States and the Soviet Union were increasing due to significant events—the construction of the Berlin Wall in 1961, the Cuban Missile Crisis of 1962, and the outbreak of war in southeast Asia. There was a sense of urgency to fulfill the dream, especially after Kennedy's assassination in November 1963. Everyone knew that spaceflight was risky, utilizing new technology and spacecraft designs that were accelerated in their development in order to reach the Moon. Recent successes increased the national self-confidence. Frustrations existed when systems didn't work properly, but no one could have predicted what happened early on in the Apollo program. On January 27, 1967, in a routine ground test in the newly built Apollo capsule, the crew of *Apollo 1* was strapped into their seats on top of the Saturn V rocket. During the test, an uncontrollable fire caused an explosion that killed all three astronauts, who were unable to escape the capsule. The program halted for an undetermined length of time to investigate the causes of the accident. The country was in shock. No one ever imagined that our astronauts could die under NASA's watchful eye.

Accident Analysis

The Apollo Command and Service Module was going through a routine test of the internal power systems. This required a fully pressurized cabin with 100 % oxygen. Three astronauts (Gus Grissom, Ed White, and Roger Chaffee) were onboard the capsule. The crew reported a fire inside of the spacecraft, and less than 20 seconds later the fire consumed the spacecraft, and the astronauts were dead (Fig. 9.1). Prior to the accident, there were multiple issues associated with the capsule design that were identified by technicians and astronauts. Gus Grissom reportedly hung a lemon on the spacecraft, making a statement about its many flaws. The day of the test, when the communications system was breaking up transmissions, Grissom said: "How are we going to get to the Moon if we can't talk between two or three buildings?"[4].

Unwilling to stop progress and impact a very compressed timeline, the testing continued. An investigation board concluded that the fire was caused by an electrical spark igniting flammable materials and exacerbated by the high pressure pure oxygen environment. To make matters worse, the astronauts couldn't exit the hatch from the inside, something that was on the list of things to fix. The analysis tells a story of human failures, not only in design flaws but also in decisions to continue testing with known problems. Crew rescue and safety issues were set aside ultimately in lieu of the bigger race to the Moon.

In looking at the accident report and analysis, it is almost unbelievable to read how the most basic issues of engineering design were violated and how none of the great minds of NASA identified the problems that led to the tragedy. Pure oxygen had been used since the start of NASA's space program. The spacecraft is designed to have a higher pressure inside than outside. In space a spacecraft can operate at low pressure because outside the vacuum of space has zero pressure. In a ground test, the spacecraft needed to operate at a higher pressure than standard atmospheric pressure, which meant that the cabin would have to be pressurized. A 100 % oxygen environment that is safe at low pressure becomes volatile at a high pressure. Despite efforts to make sure that there would be no spark inside the cabin, a spark ignited a number of flammable materials, which resulted in a deadly fire that ended the lives of the three astronauts. The inability of the astronauts to escape the capsule after the fire broke out seemed to be an unbelievable lapse of judgment in decision making. Recommendations included an engineering redesign of the module and its systems. In addition, it was recommended that management and decision procedures be revised to improve quality control.[5]

[4] Howell, Elizabeth. Space.com Contributor. 28 Aug 2012. Apollo 1: The Fatal Fire. [Internet] [cited 2015 Aug 19]. Available from: http://www.space.com/17338-apollo-1.html

[5] Moskowitz, Clara. [Internet] Space.com. 27 Jan 2012: How the Apollo 1 fire changed spaceship design forever; c2016 [cited 2016 Jun 18]. http://www.space.com/14379-apollo1-fire-space-capsule-safety-improvements.html

Fig. 9.1 *Apollo 1* module after fire (Image courtesy of NASA)

The Space Shuttle Challenger (January 28, 1986)

Historical Context

The shuttle program had been extremely successful since its first launch in 1981. A few small technical issues had been identified on the vehicle and were either redesigned or disregarded as not significant problems. In addition, certain components and technical expertise were parceled out to contractors in a similar way as during the previous space programs. This method relies on a close knit community with flawless communication processes to analyze risk. The *Apollo 1* accident was almost 20 years ago and all but forgotten in the age of the space shuttle successes. After nine successful missions, however, the public was losing interest in the program, and a new campaign to bolster enthusiasm was centered on launching the first American civilian into space. Christa McAuliffe, a schoolteacher who was selected and trained with the astronauts, was going to teach lessons from space. The mission attracted a lot of publicity but was plagued with several delays that resulted in a certain amount of pressure to launch. On a very cold morning, the *Challenger* was given a go for launch. The vehicle broke up 73 seconds after liftoff, bringing a shocking ending to the shuttle's tenth mission and the loss of seven lives onboard (Fig. 9.2).

Accident Analysis

The shuttle program was halted until an investigation could be completed by an independent commission. The data revealed the technical reason for the event—due to excessively low temperatures, two rubber O-rings that sealed joints on one of the solid rocket boosters became brittle and failed. The result was that flames leaked out of the solid booster, igniting other parts of the vehicle, including the fuel tank, causing an explosion. The non-technical reasons were just as important in the analysis. The commission focused on issues related to NASA culture and poor management decision-making processes. Scheduling pressures relating to increasing costs led to overriding the safety of the vehicle and crew and leading to the decision to launch the mission even when it was thought by some to be potentially dangerous. It was believed that NASA relied on past successes rather than engineering analysis.

Commissioner and physicist Richard Feynman thought that the communication and decision-making between managers and engineers at NASA and Morton-Thiokol, the solid rocket booster contractor, was very different in their approaches. Engineers evaluated risk statistically, whereas managers were more qualitative and at times dismissive of the detailed analyses that the engineers presented in order to move the scheduled launches along. He stated that NASA managers needed to understand the risk analysis in detail because "for a successful technology, reality must take precedence over public relations, for Nature cannot be fooled."[6]

[6] Brown, Alexander. [Internet] history.nasa.gov. Chapter 12: accidents, engineering, and history at NASA; c2015 [cited 2015 Aug 19]. Available from: http://history.nasa.gov/SP-2006-4702/chapters/chapter12.pdf

Fig. 9.2 The space shuttle Challenger explosion Jan. 28, 1986 (Image courtesy of NASA)

A House Committee on Science and Technology held hearings on the *Challenger* accident during the same year but after the commission report was made public. They interviewed members of the commission, NASA and Morton Thiokol managers, astronauts, scientists, and engineers. The committee results supported the commission findings but felt that rather than poor communication methods, the problem was inadequate technical decision-making over several years to solve problems such as the solid rocket booster joints.

NASA stopped the space shuttle program for more than 2 years while it redesigned a number of systems. Recommendations included a redesign of the solid rocket booster joints plus a change in the shuttle management structure to include astronauts in the organizational structure and have them participate in the final decision to launch a vehicle. Also included were improved safety measures and communication between all parts of the organization.

It was thought at first that the *Challenger* crew was killed instantly. After further investigation and the recovery of the crew cabin at the bottom of the ocean, it was discovered that the crew most likely lived more than 2 minutes past the separation from the solid rocket boosters. The only thing that could be certain was that the impact with the ocean would have been fatal. The crew cabin was intact most likely until impact with the ocean. By analysis of flight data, it was determined that the *Challenger* broke apart at 48,000 feet above the ocean, arced to 65,000 feet before beginning its flight downward. Evidence showed that the reserve oxygen packs had been turned on, and the g forces were survivable. All seven astronauts were most likely alive and conscious during most of those terrifying last 2 minutes and 45 seconds, trying to save themselves as the cabin plunged into the ocean. This part of the story needed to be told. There was no point in sanitizing the event. However, the official NASA report concluded, "The cause of death of the *Challenger* astronauts cannot be positively determined."[7]

It is important to reflect on these facts to understand the choices that NASA made concerning crew safety. An emergency escape module with a parachute system was proposed along with other escape systems, but it was thought that none of the systems would have been capable of saving the *Challenger* crew. In addition, escape systems added significant weight to the launch vehicle and affected the shuttle's capacity to carry payloads, so they weren't considered viable options. The reality was that there were no options in place for the survival of crew members during powered ascent.

[7] Harwood, William. [Internet] Space-shuttle.com. The fate of challenger's crew; c2010 [cited 2015 Aug 21]. Available from: http://www.space-shuttle.com/challenger1.htm

The Space Shuttle Columbia (February 1, 2003)

Historical Context

The shuttle program made a successful return after the *Challenger* accident. Some systems were upgraded, and significant changes were made in the organizational structure and decision-making process. Some technical issues that existed since the first shuttle flight, though, were never addressed because of the significant cost to change the design or the expected low risk of failure.

Columbia was the first orbiter vehicle in the space shuttle fleet with its first launch on April 12, 1981. It was in service for over 22 years and had completed 27 missions before nearing the end of its 28th mission, STS-107 on February 1, 2003.[8] The *Columbia* flight was delayed for a number of times over a 2-year period. While re-entering the atmosphere over Texas and only a few minutes to landing at Kennedy Space Center, the *Columbia* broke apart, killing all seven crew members aboard. The *Columbia* was the second tragedy in the space shuttle program following the *Challenger* explosion.

Accident Analysis

The shuttle program was halted until an investigation could be completed by an independent commission. The investigation revealed that the re-entry disaster was caused by a problem that occurred at launch almost two weeks before. At about 80 seconds into the launch, a piece of foam insulation broke off from the shuttle's propellant tank and damaged the edge of the left wing of the orbiter vehicle.[9] This was not an unusual event. Even on the first space shuttle flight, a small piece of foam came off of the external tank and hit the tiles on the underside of the orbiter vehicle. The ceramic tiles, or Thermal Protection System (TPS), used on the underside of the orbiter vehicle, are heat resistant but fragile and can break or chip easily. The main concern during the first shuttle flight was whether the chipped or fragmented tiles compromised the re-entry capability of the tiles as a heat shield. It was thought that a loss of specific tiles or damage in the leading edge of a wing on the orbiter

---

[8] Howell, Elizabeth. [Internet] Space.com. 16 Jan 2013: Columbia: first shuttle in space; c2016 [cited 2016 Jun 19]. Available from: https://www.nasaspaceflight.com/2011/02/space-shuttle-columbia-a-new-beginning-and-vision/

[9] History.com Staff. 2010. [Internet] history.com. Columbia disaster; c2015 [cited 2015 Aug 19]. Available from: http://www.history.com/topics/columbia-disaster

would allow severe heat to enter the wing cavity and produce a "zipper effect," causing other tiles to be removed by increased aerodynamic pressure.

On the first shuttle flight and a number of subsequent flights the damage incurred wasn't significant. The same situation existed for the doomed *Columbia* flight. In an area of severe wind buffeting, fuel tank foam debris hit the underside of the orbiter, damaging some tiles. Management decided not to tell the crew about it because there was little to be done to fix the problem. The vehicle was not equipped to dock with the International Space Station, and a maneuver to a higher altitude would have used up a great amount of fuel. It was unclear whether an EVA would have revealed important information without significant risk to the astronauts. It was also decided by the shuttle program manager that taking satellite imagery to inspect the damage was not going to be helpful. There was no tile repair kit that was workable at that time.

The investigation board revealed that it actually would have been possible for the crew to repair the tile damage or for the crew to be rescued from the *Columbia* by the launch of the shuttle *Atlantis*, which could have helped repair the tiles or taken the crew onboard.

The space shuttle program was grounded for over 2 years to implement some of the recommendations of the board. Many of the same issues, such as organizational causes and flawed decision-making processes instrumental in the *Challenger* explosion, were still to be blamed for the *Columbia* disaster. Some of the reasons why it happened were also similar—a need to gain approval and publicity for public enthusiasm along with the lack of national vision and commitment to the space program, decreasing resources, changing priorities, and schedule pressures.

When the shuttle program resumed, the ability to take pictures of the underside of the orbiter for visual analysis of affected tiles and repair tiles using a kit and an EVA had been introduced. These items were tested but never required for the remainder of the shuttle program.

Lessons Learned from NASA Space Disasters

There are some obvious similarities between all three NASA investigation recommendations, particularly in the process of decision making. Communication between managers and engineers and among contractors in any business can be challenging. Most poorly constructed decisions don't result in a disastrous loss of life, but NASA's pervasive culture of success or even arrogance contributed to questionable decisions concerning crew safety and risk. All three investigations recognized a growing separation of management and engineering, which is problematic if managers do not understand the technical aspects of risk. A number of technical issues were not addressed for long periods of time, despite reports from scientists and engineers, both at NASA and from contractors.

The fact that these issues were not properly addressed on a regular basis contributed to each fatal event. Why they were not addressed had patterns as well—addressing technical issues in depth and implementing solutions could put a program at risk due to financial requirements and schedule delays. It was thought that engineers wanted everything perfect, and if each item got the detailed attention and time requested, no one would ever have been launched into space. With the managers and engineers not communicating effectively, the concerns could easily be overlooked and not factored into important decisions.

All of the accidents were caused by issues that were already identified as potentially dangerous, and yet no solutions were implemented because no adverse events had occurred up until that point for those issues. Engineering problems that had not caused failures were put into the category of acceptable systems rather than a database of issues of concern. Lack of disaster was accepted as success. A sigh of relief when a mission was completed translated into no further testing or analysis on the items in question. A blind spot was created with respect to a number of technical issues that merited additional testing. What is disturbing is that, after decades of NASA space missions, the Columbia Accident Investigation Board (CAIB) reported that the NASA culture hadn't changed much since the *Challenger* disaster and stated that the space agency lacks "effective checks and balances, does not have an independent safety program and has not demonstrated the characteristics of a learning organization." The report continues to say, "these repeating patterns mean that flawed practices embedded in NASA's organizational system continued for 20 years and made substantial contributions to both accidents."[10]

It was thought that NASA might never get this aspect of organizational communication right and maybe were destined to face another tragic event. One of the most powerful conclusions in the report was that NASA had recreated an attitude and culture that they vowed would never return—one in which "engineers had to produce evidence that the system was unsafe rather than prove it was safe."[11] It was clear that progress had to be made in terms of safety and communication going forward.

[10] Lessons unlearned: NASA is to blame for the Columbia disaster. Pittsburgh Post—Gazette. 2003 Sep 02.

[11] Report blames flawed NASA culture for tragedy; in broad indictment of practices, Shuttle panel says safety suffered. The Washington Post. 2003 Aug 27.

The space shuttle program did retire successfully in 2011 after returning to flight in 2005. One significant change that demonstrated NASA's commitment to safety going forward was the establishment of the NASA Engineering Safety Center and a Chief Safety Officer for the space shuttle program. The CAIB recommended several changes, and although several of its members thought the shuttle should be retired, it was not the board's official recommendation. The report did describe the shuttle program as "a complex and risky system."[12] The decision was made in 2004 to retire the shuttle and rely on the Soyuz vehicles to transfer US crews to the ISS.

NASA had planned for the shuttle to retire several times over the years but kept it going due to its success overall and the delayed status for a replacement. The shuttle would still be needed for ferrying astronauts and supplies to the International Space Station until 2020. The CAIB recommended a vehicle recertification on all levels to ensure flight safety if the shuttle was to be used beyond 2010. The 2010 date was chosen because the ISS was supposed to be completed by that date, and the shuttle use beyond that date was not anticipated. Soon after, NASA set 2010 as the retirement date.[13] The shuttle was retired shortly after in 2011.

It is thought that, despite past evidence to the contrary, NASA is the safest organization to oversee space exploration. With the emergence of private enterprise, the safety of space exploration is being questioned. It is a valid question, considering the challenges that NASA has had over the past several decades. However, there is some benefit to having a smaller organization, which has control over all systems and mission planning, but being small doesn't necessarily guarantee success. Lessons can be learned from these fatal events.

SpaceShipTwo Crash (October 31, 2014)

Historical Context

Burt Rutan and Sir Richard Branson joined forces to design and build the first commercial spacecraft. Burt Rutan had already launched SpaceShipOne, the first private manned rocket into space. Virgin Galactic was founded in 2005 with the goal of developing a space tourism industry. It designed and built

[12] Lessons unlearned: NASA is to blame for the Columbia disaster. Pittsburgh Post—Gazette. 2003 Sep 02.

[13] Day, Dwayne A. The decision to retire the space shuttle. The Space Review. 2011 July 18.

SpaceShipOne and a space hub in New Mexico for Virgin Galactic flights. An explosion in the testing of one of their rocket engines in July 2007 killed three technicians, injuring other employees.[14]

Undaunted by this event, the company moved forward and developed SpaceShipTwo, a six-passenger space plane with two pilots. The craft would be carried underneath a carrier vehicle to a high altitude, and then released to start its own rocket engine and fly into space beyond Earth's atmosphere. On a test flight on the last day of October in 2014, the vehicle manned by the two pilots was carried by the WhiteKnightTwo vehicle to an altitude of about 45,000 feet over the California's Mojave Desert and released. The rocket engines started, and the vehicle started to accelerate to a higher altitude when it exploded, splitting into pieces and falling back to Earth, killing one of the pilots (Fig. 9.2).[15]

Accident Analysis

The National Transportation Safety Board (NTSB) determined that the crash of SpaceShipTwo was caused by a combination of human error and inadequate safety procedures. The pilot who died in the crash prematurely unlocked what was called the "feathering system," a device that is used to reduce speed on descent. A possible reason for the early deployment of the device was to prevent the flight from being aborted. Because this event wasn't anticipated, there were no built-in safeguards to prevent it. The NTSB suggested some new safety procedures, including a modification to the feathering lock system. Sir Richard Branson was still resolved to go forward with commercial space tourism.[16]

SpaceX Explosion (June 28, 2015)

Historical Context

SpaceX, founded in 2002 by entrepreneur Elon Musk, produced the first private spacecraft to dock with the ISS and deliver supplies under a NASA resupply contract. The company wants to replace the Russian Soyuz rocket as a transport for American astronauts to the ISS, a necessity after the space

[14] Chang, Alicia. Virgin Galactic keeps low profile after explosion. USA Today. 2007 Aug 26.

[15] Mail FS. 03 Nov 2014. Moment Virgin Galactic spaceship exploded at 45,000 ft. Daily Mail. [London (UK)]: 13:17.

[16] Gajanan M. Virgin galactic crash: Co-pilot unlocked braking system too early, inquiry finds. The Guardian. 2015 Jul 29; Sect. 11.

shuttle retired in 2011. It also wants to develop the technologies necessary for a reusable launch system, reducing space transportation costs considerably. Long-term goals for SpaceX include the colonization of Mars.

On June 28, 2015, one of the SpaceX Falcon 9 rockets experienced an overpressure event just over 2 minutes into flight, resulting in a loss of the mission.[17] The rocket was loaded with food and supplies for the ISS and was the worst failure in the company's history, although no lives were lost.

Accident Analysis

Thus far, the investigation has identified a single piece of hardware that failed, causing the overpressure event. The piece that broke was a steel strut, 2 feet long, manufactured by a supplier. The strut was holding down one of many helium bottles on the rocket's second stage.[18] Interestingly, Elon Musk believed that SpaceX employees may have become too relaxed as a result of their numerous successes in recent years, causing quality control to suffer. "Most people at the company today have only ever seen success," Musk said. "When you've only ever seen success, you don't fear failure quite as much."[19] Musk said that parachutes onboard the cargo ship could be deployed by software in an emergency, saving its cargo. He said this fix would be added to future missions. Delays are expected to be several months in scheduling and costing millions of dollars.

Lessons Learned from Commercial Space Disasters

A constant throughout almost all of the fatal accidents related to space travel is either human error or insufficient attention to safety. An overseeing safety officer and rigorous quality control was recommended in these accidents. Complacency was mentioned by Elon Musk as a possible reason for the lapse in focus on safety. This is very similar to the NASA culture referred to in all fatal accidents. Maybe it is just human nature to drop your guard when you have experienced so many successes. And if so, some procedures to prevent this from happening need to be put in place.

[17] Petersen M. SpaceX loss blamed on faulty strut; the snapping of a steel rod caused the rocket to explode after liftoff, says founder Elon Musk. Los Angeles Times. 2015 Jul 21.

[18] Strut may be cause of SpaceX accident. Chicago Tribune. 2015 Jul 21; Sect. 11.

[19] Petersen M. SpaceX founder Elon Musk blames rocket failure on shoddy part. Los Angeles Times. 2015 Jul 20.

10

New Technology and Deep Space

Keywords 100 Year Starship (100YSS) • Ad Astra Rocket Company • Advanced Life Support Research • Aerojet Rocketdyne • CERN (the European Organization for Nuclear Research) • Defense Advanced Research Projects Agency (DARPA) • Dr. Mae Jamison • Dr. Ray Wheeler • Enhanced Power Technology • Environmental Control and Life Support System • European Space Radiation Superconducting Shield (SR2S) project • Exploration Research and Technology Program • Game Changing Development (GCD) program • Habitat Technology • in-situ resource utilization • *Interstellar* • Ion propulsion • ISS • John Grunsfeld • Lanetra Tate • NASA Evolutionary Xenon Thruster (NEXT) ion thruster • NASA's Mars Reconnaissance Orbiter • NextSTEP habitat projects • Paul Sabatier • Radiation Protection Technology • Roberto Battiston • Sabatier System • Solar array development • Solar arrays • Solar Sails • *The Martian* • Vasimr engine • William Gerstenmaier

> "To explore strange new worlds, to seek out new life and new civilizations, to boldly go where no man has gone before."
>
> *Star Trek*

The introductory line from the "Star Trek" TV show and movies is bold and inspiring. Actually, part of this speech, delivered by Captain James T. Kirk, can be found in a 1958 document entitled "Introduction to Outer Space:": "…the compelling urge of man to explore and to discover, the thrust of curiosity that leads men *to try to go where no one has gone before.*"[1]

[1] History.NASA.gov. President's Science Advisory Committee, "Introduction to Outer Space," [Internet] History.NASA.gov; March 26, 1958 [cited 2016 Mar 10]. Available from: http://history.nasa.gov/sputnik/16.html

L. Dawson, *The Politics and Perils of Space Exploration*, Springer Praxis Books, DOI 10.1007/978-3-319-38813-7_10

The document was produced by President Eisenhower's Science Advisory Committee as a brief non-technical promotional document outlining the reasons why America should pursue a space program. Decades later, we can still relate to the same factors identified in this report as we look to further human exploration into deep space. First, the desire to explore the unknown seems to be as popular now as it was over 50 years ago. Second, having a military advantage in the use of outer space remains an issue with all countries who are major players in world politics. Third, the national prestige that comes as a result of demonstrating the capability to launch space vehicles or satellites is still a prominent reason for many countries to develop a space program.

Finally, it is still accepted that the scientific observations and experiments away from Earth will expand our knowledge of the universe. The report included a brief discussion about the development of space technology being at the cutting edge of our capability and that uncertainties and failures are expected in its implementation. Today, we have a realistic perspective on these issues, and having lost several lives in the space program, we approach the next adventures cautiously, maybe too cautiously for many.[2]

Almost 60 years after this report was released, we still view humans traveling in futuristic spaceships at warp speeds (or even fast speeds) to far-off planets as science fiction. That doesn't stop us from dreaming about the possibilities that await us, and we are always hoping it happens in our lifetime. We support the notions in current books and in movies such as *The Martian*, which portrays an astronaut marooned on Mars, using his scientific knowledge and creativity to survive. The movie *Interstellar* projects a bleak future for Earth when NASA sends a former astronaut through a wormhole to the other side of the galaxy to find another habitable planet to colonize. Our fascination with and interest in futuristic concepts and travel through space has been a constant for decades. Our own space program hasn't met this desire yet, and other than our missions to the Moon, manned missions have been stuck in low-Earth orbit for decades. We are anxious to break away from Earth and explore deep space, but we are limited by technological developments. This chapter address the technological challenges that keep Star Trek fiction instead of reality.

---

[2] History.NASA.gov. President's Science Advisory Committee, "Introduction to Outer Space," [Internet] History.NASA.gov; March 26, 1958. [cited 2016 Mar 10]. Available from: http://history.nasa.gov/sputnik/16.html

Deep Space Exploration Technological Challenges

The only spacecraft that have traveled to the outer reaches of our Solar System are unmanned vehicles that no longer communicate with Earth. In August 2012, transmitted data provided evidence of NASA's *Voyager 1* (launched in 1977) as the first manmade object to go past the heliosphere, the magnetic boundary at the outer edges of the Solar System.[3] We will never hear from Voyager again, so we don't know what it will encounter because it is beyond our capabilities to hear its transmissions.

There is interest in human travel to other planets and beyond, but we have to meet the technological challenges that are delaying this travel. First of all, these destinations involve incredibly long distances, and our current methods of propulsion would require long periods of time (many, many years) to get to even the nearest of these locations, so we need to develop newer, faster propulsion methods. We don't have the ability to travel "warp" speed (faster than the speed of light), and so each mission requires that humans be transported with water and food or the ability to make water and food for very long periods of time. Add to that the limitations in communication and rescue, and the missions become extremely dangerous.

Currently, NASA is planning on manned missions to an asteroid by 2025 and Mars in the 2030s.[4] Mars is about 36 million miles from Earth (about 58 million km). Using our current methods of propulsion, similar to what was used to go to the Moon, Mars is reachable within 1.6 years at Earth's closest approach. However, if we wanted to visit the nearest star system, Proxima Centauri, the distance increases to over 21 trillion miles (almost 40 trillion km) and lies well beyond our imagination and technological development.[5] Voyager travels at 3.6 Astronomical Units (AU) per year. (AU is the average distance between Earth and the Sun, which is about 93 million miles or 150 million km.) At this speed, it would take over 75,000 years to reach Proxima Centauri.[6]

---

[3] Landau, Elizabeth. CNN news. [Internet] cnn.com; Voyager 1 becomes first human-made object to leave solar system. Oct 2, 2013; [cited 2016 Mar 10]. Available from: http://www.cnn.com/2013/09/12/tech/innovation/voyager-solar-system/

[4] NASA.gov. [Internet] NASA.gov; 2015 [cited 2016 Mar 11]. Available from: http://www.nasa.gov/content/nasas-journey-to-mars

[5] Science channel. [Internet] sciencechannel.com; 10 technology innovations needed for deep space exploration. 2015 [cited 2016 Mar 10]. Available from: http://www.sciencechannel.com/topics/aliens-space/10-technology-innovations-needed-for-deep-space-exploration/

[6] NASA.gov. Voyager forges a new frontier. 2011 [Internet] NASA.gov; [cited 2016 Mar 11]. Available from: http://www.nasa.gov/missions/deepspace/voyager_prt.htm

The issues of speed and distance affect an unmanned spacecraft. If human travel is considered, as previously discussed, we have to address the other complications and issues of space travel that affect humans, including protection against long-term exposure to radiation and counteracting long term effects of microgravity. Basic requirements of food and water as well as communication with Earth on a regular basis are huge challenges.

NASA is in the early stages of development for a number of new technologies that will be useful for deep space travel. In 2012, the Defense Advanced Research Projects Agency (DARPA), the Pentagon's research and development branch, awarded $500,000 in seed money to the Dorothy Jamison Foundation for Excellence to form the 100 Year Starship (100YSS), an independent, non-government initiative that will call on experts from a variety of fields, artists to engineers, in order to develop the capabilities for human interstellar flight "as soon as possible, and definitely within the next 100 years." Dr. Mae Jamison, a former NASA astronaut and the creator and leader of the organization granted the winning proposal, said: "Yes, it can be done. Our current technology arc is sufficient …100 Year Starship is about building the tools we need to travel to another star system in the next 100 years." The first year of the ambitious project will involve searching for investors, establishing membership opportunities, encouraging public participation in research projects, and developing the visions for interstellar exploration.[7]

The 100YSS is a long range project created to ensure that humans reach the stars within a 100 years. The seed funding is only sufficient to provide some kind of business plan for the future and to support research toward the creation of a starship within 100 years.

In looking shorter term, NASA scientists and others are working on technological developments that can meet the challenges of deep space travel. In 2015, as part of NASA's Space Technology Research Grants Program, eight academic proposals were awarded with grants of approximately $200,000 to study technologies currently at an early stage of development, such as solar cells that operate at high temperatures, extremely strong and lightweight structures, and synthetic biology applications for recycling human waste.[8]

In addition, at about the same time, NASA funded new space technologies under the NextSTEP program (Next Space Technologies for Exploration

[7] David, Leonard. NASA's 100-year starship project sets sights on interstellar travel. 23 Mar 2011. [Internet] [cited 2016 Mar 11]. Available from: http://www.space.com/11200-nasa-100-year-starship-interstellar-travel.html

[8] Buck, Joshua. NASA awards grants for technologies that could transform space exploration. 14 Aug 2015. [Internet] NASA.gov; [cited 2016 Mar 13]. Available from: http://www.nasa.gov/press-release/nasa-awards-grants-for-technologies-that-could-transform-space-exploration

Partnerships). This program established partnerships with ten aerospace companies and one university to develop expertise the three major areas: advanced electric propulsion, human habitation, and small satellites.

"Commercial partners were selected for their technical ability to mature key technologies and their commitment to the potential applications both for government and private-sector uses," William Gerstenmaier, associate administrator for Human Exploration and Operations at NASA headquarter said in a statement. "This work ultimately will inform the strategy to move human presence further into the Solar System."[9]

This chapter investigates a few of the technologies that are key to making manned deep space travel possible.

Methods of Propulsion

One of the limiting factors for manned spaceflight is the current capability of our propulsions systems. The limits of our current systems are available speed and amount and type of fuel required to transport. The most difficult part of a space journey is escaping Earth's gravitational pull. Huge multi-stage rockets were used to project a small lander and human capsule to the Moon. Each stage was used to push the vehicle faster and higher. When depleted, each stage fell away, and the lighter vehicle continued on its trajectory. Right now, rockets are our only option for getting vehicles into orbit or for travel beyond Earth. The amount of fuel and the oversized rockets themselves are very expensive. Once a spacecraft is out of the planet's gravitational pull, there are other means of propulsion that are efficient but not particularly fast.

Ion Propulsion

Ion propulsion is one method used frequently in science fiction for manned flight, but in reality it is currently used to keep satellites in their proper orbits (station-keeping) and to send probes on long journeys into deep space. NASA began researching the use of ion propulsion as early as the 1950s. In 1998, ion propulsion was successfully used as the main propulsion system for the *Deep Space 1* mission.

Deep Space 1 (1998–2001) was the first spacecraft to use almost entirely ion propulsion to travel over 163 million miles and fly by asteroid Braille and

[9] Venturespring. NASA funds 12 deep-space exploration technologies. 2 Apr 2015. [Internet] venture-spring.com; [cited 2016 Mar 14]. Available from: http://venture-spring.com/news/?p=64

comet Borelly. Currently, ion thrusters are used to keep more than 100 geosynchronous Earth orbit communication satellites in their desired locations. In addition, ion thrusters are propelling the Dawn spacecraft (launched 2007) to travel inside the Asteroid Belt between Mars and Jupiter and orbit two celestial bodies called Vesta and Ceres.[10]

Ion thrusters generate only a small amount of thrust. They have high specific impulses, which represent the ratio of thrust to the rate of propellant consumption. This means that they need less propellant (5–10 times less fuel) than a more traditional chemical propulsion system. In addition, they can achieve very high speeds (theorized up to 200,000 mph) but only after operating a very long time prior to that high speed. Acceleration is small, but if sustained will become quite large. The ion thruster ionizes propellant, producing ions by adding or removing electrons. The gas produced, called plasma, contains both positive ions and negative electrons of equal proportions and, therefore, has no resulting charge. The plasma acts like any gas but is affected by electric and magnetic fields, much like the material inside fluorescent light bulbs. The most common propellant used in this application is xenon, which is easily ionized and inert, with a high storage density that suits it well for storing on spacecraft.[11]

Future missions require faster propulsion than what the ion thruster currently provides within a reasonable timeframe. One of the companies that NASA selected in 2015 to develop technology using a NextSTEP grant was the Ad Astra Rocket Company in Texas. They are developing the Vasimr engine, which uses plasma as a propellant. This engine has been advertised as being able to propel a spacecraft to Mars in 39 days. Over a 3-year period, starting in 2015, NASA will award Ad Astra with $10 million to develop a flight-ready version of the successful prototype.[12]

A competing design, also funded by the NextSTEP program, is a small thruster called "X3" (XR-100), developed by Alec Gallimore of the University of Michigan. The XR-100 also uses plasma ejected out of the back of the spacecraft as propellant, which accelerates the vehicle to very high speeds. In this design, a current of electrons travels through a circular channel, the spiraling motion building an electric field that pulls the gas ions out of the exhaust at the end of the channel.[13]

---

[10] NASA.gov. NASA—ion propulsion. 11 Jan 2016. [Internet] NASA.gov; [cited 2016 Mar 13]. Available from: http://www.nasa.gov/centers/glenn/about/fs21grc.html

[11] Hicks, Kenneth. Ion propulsion might carry spacecraft far, wide. 17 Jan 2016. Columbus Dispatch. News—Science: 3H.

[12] Aa Cihan News Agency. NASA selects companies to develop super-fast deep space engine. April 2, 2015.

[13] Mirror Publications. New thruster to propel future Mars mission. 20 Feb 2016.

Fig. 10.1 Artist's concept of the advanced Solar Electric Propulsion System (SEP). (Image courtesy of NASA)

NASA is also involved in the development of two different ion thrusters: the NASA Evolutionary Xenon Thruster (NEXT) and the Annular Engine. NEXT is a high-power design to reduce mission cost and trip time. NASA Glenn recently awarded a contract to Aerojet Rocketdyne to fabricate two NEXT flight systems (thrusters and power processors) for use on future NASA science missions and to develop the NEXT technology to produce increased power and thrust-to-power to be used for a broad range of commercial, NASA, and defense applications (Fig. 10.1).[14]

Solar Sails

One of the more unusual although promising ideas is to equip spacecraft with giant sails designed to capture solar energy and use it for propulsion. This would be possible because light is made up of tiny particles called photons, which in some ways behave like atomic particles. When these photons strike a mirror-like surface, they are reflected straight back. In this process, they transmit their momentum to the surface, pushing it forward ever so slightly.

[14] NASA.gov. NASA—ion propulsion. 11 Jan 2016. [Internet] NASA.gov; [cited 2016 Mar 13]. Available from: http://www.nasa.gov/centers/glenn/about/fs21grc.html

Fig. 10.2 NanoSail-D is made of extremely lightweight gossamer fabric designed to glide into space (Image courtesy of NASA)

One advantage of a solar sail would be similar to that of an ion thruster; it would cause a spacecraft to steadily accelerate, so that the craft eventually would reach an extremely high rate of speed. One drawback of solar sails is that a sail would have to be many times larger than that spacecraft in order to provide the force necessary to propel the spacecraft forward. These types of sails might actually have to be built in space (Fig. 10.2).[15]

Habitat Technology

Another major challenge for deep space manned flight is being able to set up habitats, most likely prior to a crew actually arriving on the planet or celestial body. NextSTEP habitat projects are focused on developing modules to add onto the Orion space capsule, which is designed to support a four-person crew for up to 3 weeks in deep space. NASA wants to increase that capability to 60 days, with the potential for increased capability for a mission to Mars. Modular systems are the most popular concept for easy assembly. The seven NextSTEP habitat projects will have initial performance periods of up to 12 months, with a funding of $400,000 to $1 million for the study and development efforts. Some of the selected companies are Bigelow Aerospace, the

[15] Wall, Mike. Space.com Staff Writer. 27 May 2014. Are solar sails the future of space travel. [Internet] [cited 2016 March 14]. Available from: http://www.space.com/26011-solar-sail-tech-space-exploration.html

Fig. 10.3 Artist's concept of a deep space habitat. (Image courtesy of NASA)

Boeing Company, Lockheed Martin Space Systems, and Orbital Technologies Corporation (Fig. 10.3).[16]

The Orion spacecraft that will take astronauts to Mars has a diameter close to 16 feet, not a lot of space when you consider the journey to Mars will take at least 6 months. The astronauts will need a larger place to live with private quarters and exercise equipment. NASA envisions the Orion capsule linking up to a habitation module in space, which has yet to be designed. A government spending bill, part of the Exploration Research and Development funding at the end of 2015, allocated at least $55 million to develop a habitation module for deep space exploration, including a prototype ready by 2018. This is the time when NASA plans to test its new space habitat around the Moon in the 2020s before sending it to Mars in the 2030s. The government clearly designated the focus to be on the habitat module, an important component for future long-term missions (Fig. 10.3).[17]

[16] Foust, Jeff. Spending bill to accelerate NASA habitation module work. 28 Dec 2015. [Internet] [cited 2016 March 14]. Available from: http://spacenews.com/spending-bill-to-accelerate-nasa-habitation-module-work/

[17] Foust, Jeff. Spending bill to accelerate NASA habitation module work. 28 Dec 2015. [Internet] [cited 2016 March 14]. Available from: http://spacenews.com/spending-bill-to-accelerate-nasa-habitation-module-work/

Radiation Protection Technology

We have previously discussed the dangers of long-term exposure to radiation. It's important to look at advancements in this area and what new technologies exist for this challenge for extended manned spaceflight. Scientists at CERN (the European Organization for Nuclear Research) have announced they are working on a solution to this very problem. In collaboration with the European Space Radiation Superconducting Shield (SR2S) project, CERN is developing a superconducting magnetic shield that can protect a spacecraft and its crew from cosmic rays during deep-space missions. The shield will generate a field strong enough to stop the harmful rays from penetrating spacecraft and harming crewmembers and equipment. The cosmic radiation issues are estimated to be solved within 3 years. The goal of the SR2S project is to create a magnetic field that is 3000 times stronger than Earth's own magnetic field, with a 10-meter diameter that would protect crews inside or just outside a spacecraft.[18]

In a press release, Roberto Battiston, project coordinator of SR2S said, "This situation is critical. According to our present knowledge, only a very small fraction of NASA's active astronauts are suitable to stay on the ISS for a 1-year mission regardless of the fact that the exposure to radiation is two times less than the exposure during deep space travel. The next exploration challenges, deep space travel to near Earth asteroids and long-duration stay on Mars and on the Moon, require an effective way to actively shield astronauts."[19]

Communication Technology

The ability to communicate with Earth is essential for a long duration, deep-space manned mission. The transmission of data over long distances in a timely fashion is also a major challenge. "If you can't communicate with the ship, then you don't know what the results are of your mission," Andreas

[18] Dockrill, Peter. Science Alert. Scientists are developing a shield to protect astronauts form cosmic radiation. 6 Aug 2015. [Internet] [cited 2016 March 14]. Available from: http://www.sciencealert.com/scientists-are-developing-a-magnetic-shield-to-protect-astronauts-from-cosmic-radiation

[19] Dockrill, Peter. Science Alert. Scientists are developing a shield to protect astronauts form cosmic radiation. 6 Aug 2015. [Internet] [cited 2016 March 14]. Available from: http://www.sciencealert.com/scientists-are-developing-a-magnetic-shield-to-protect-astronauts-from-cosmic-radiation

Tziolas, a former research fellow at NASA who now heads Project Icarus, a private-sector effort to develop interstellar technology.[20]

Early space communications were radio-based, with significant time delays. These methods are challenged by needs for significantly higher data rates with less mass and power, also critical to any spacecraft. Newer technology has almost eliminated delays in telecommunications on Earth, but they still occur for long-distance space transmissions and can be significant to transmit to Mars (over 30 minutes). This can impact mission objectives and the safety of the crew. NASA is currently funding the Laser Communications Relay demonstration that uses laser beams to transfer data between spacecraft and relay stations on Earth at 10–100 times the current speeds and several times faster than the average broadband Internet connect. Images would take longer to transmit, going from Mars to Earth in about five minutes.[21] The wavelength of laser light is orders of magnitude shorter than radio waves, which makes the energy more focused and less wasteful as it travels through space. In addition, the shorter wavelength allows for significantly more bandwidth available for transmissions. High-resolution measurements and image transmission will help looking at Earth's own climate and environmental changes as well as studying new planets. The laser communications system will be tested in orbit as a commercial satellite payload in 2019.[22]

Enhanced Power Technology

Future deep space missions will require solar arrays that can operate under very low temperature conditions and exposed to high radiation. The development of a new generation of solar power technology to address these issues is critical to improve mission performance and increase power output as well as solar array life in the harsh environment of deep space.

NASA's Game Changing Development (GCD) program has selected four proposals to develop solar array technologies that will aid spacecraft in exploring destinations well beyond low-Earth orbit, including Mars: "These awards

[20] Science channel. [Internet] sciencechannel.com; 10 technology innovations needed for deep space exploration. 2015 [cited 2016 Mar 15]. Available from: http://www.sciencechannel.com/topics/aliens-space/9-super-high-speed-optical-communication/

[21] Science channel. [Internet] sciencechannel.com; 10 technology innovations needed for deep space exploration. 2015 [cited 2016 Mar 15]. Available from: http://www.sciencechannel.com/topics/aliens-space/9-super-high-speed-optical-communication/

[22] NASA.gov. Laser communications relay demonstration (LCRD). 2015 [Internet] NASA.gov; [cited 2016 Mar 15]. Available from: http://www.nasa.gov/mission_pages/tdm/lcrd/lcrd_overview.html#.Vuh60uIrK70

will greatly enhance our ability to further develop and enhance LILT [low-intensity low temperature] performance by employing new solar cell designs," said Lanetra Tate, the GCD program executive in NASA's Space Technology Mission Directorate. "The ultimate goal of increasing end of life performance and enhanced space power applications will greatly impact how we execute extended missions, especially to the outer planets." [23]

Four proposals for solar array development (from John Hopkins University, the Boeing Company, NASA's Jet Propulsion Laboratory, and ATK Space Systems) were selected for contract negotiations, focusing on the development of solar arrays for extreme environments. Initial awards are as much as $400,000, funding for about 9 months of system design, component testing, and analysis, at which time NASA would award additional money ($1.25 million) to develop and test the hardware. A winner would be selected to develop the final design.[24]

Food Crops for Deep Space Applications

One critical capability for human survival in long-distance manned missions in deep space is the ability to produce food. NASA is preparing for its mission to Mars and the necessity of growing food crops in controlled environments for long-duration manned missions. Dr. Ray Wheeler is doing research into growing food in a chamber.

Dr. Wheeler, a NASA plant physiologist, talks about the challenge of growing food on a planet with an extreme environment: "*The Martian* movie and book conveyed a lot of issues regarding growing food and surviving on a planet far from Earth. It's brought plants back into the equation"[25] (Fig. 10.4).

Fictional astronaut Mark Watney from the movie *The Martian* and Dr. Wheeler are both botanists, although one is fictional. Wheeler, however, is

---

[23] NASA.gov. NASA selects proposals to build better solar technologies for deep space missions. RELEASE 16-032.

14 Mar 2016. [Internet] NASA.gov; [cited 2016 Mar 15]. Available from: http://www.nasa.gov/press-release/nasa-selects-proposals-to-build-better-solar-technologies-for-deep-space-missions

[24] NASA.gov. NASA selects proposals to build better solar technologies for deep space missions. RELEASE 16-032.

14 Mar 2016. [Internet] NASA.gov; [cited 2016 Mar 15]. Available from: http://www.nasa.gov/press-release/nasa-selects-proposals-to-build-better-solar-technologies-for-deep-space-missions

[25] Herridge, Linda. NASA.gov. NASA plant researchers explore question of deep-space food crops. 17 Feb 2016. [Internet] NASA.gov; [cited 2016 Mar 15]. Available from: http://www.nasa.gov/feature/nasa-plant-researchers-explore-question-of-deep-space-food-crops

Fig. 10.4 A variety of red potatoes, called Norland, were grown in the Biomass Production Chamber inside Hangar L at Cape Canaveral Air Force Station in Florida during a research study in 1992 (Image courtesy of NASA)

the lead for Advanced Life Support Research activities in the Exploration Research and Technology Program at Kennedy Space Center, working on real plant research. Wheeler and his colleagues, including plant scientists, have been studying ways to grow safe, fresh food crops efficiently off Earth. Most recently, astronauts on the International Space Station harvested and ate a variety of red romaine lettuce that they activated and grew in a plant growth system called Veggie. Wheeler was among the plant scientists and others who helped get the Veggie unit tested and certified for use on the ISS. The unit requires a chamber with low power requirements and low mass.[26]

Water, light, and soil, along with some nutrients, are also necessary for growing potatoes or sweet potatoes, as an example. Other crops such as wheat and soybeans would also work in the chamber, providing carbohydrates and protein. In addition, potatoes are tubers, storing their edible biomass in underground structures. Wheeler said potatoes could produce twice the amount of food as some seed crops when given equivalent light. After salad crops that are now being studied, they are the next category of minimally processed food crops and could be consumed raw.

"You could begin to grow potatoes, wheat and soybeans, things like that, and along with the salad crops, you could provide more of a complete diet.

[26] Herridge, Linda. NASA plant researchers explore question of deep-space food crops. 17 Feb 2016. [Internet] NASA.gov; [cited 2016 Mar 15]. Available from: http://www.nasa.gov/feature/nasa-plant-researchers-explore-question-of-deep-space-food-crops

Providing food is a complex issue," Wheeler said. "We have to think about nutritional issues, what's acceptable and what tastes good. If nobody wants to eat it, that won't work."[27]

In the movie, the character chooses to use the regolith, or Martian soil, to grow the plants. In reality, the soil on Mars is essentially broken rock material, and lacks most of the nutrients needed to sustain plant growth. Much of what Wheeler did in his potato studies involved growing the plants in shallow, tilted trays using a hydroponic system.

"With potatoes, it was a little bit more interesting in the sense that you can't use systems that require a lot of standing or deep water—potatoes don't like to be submerged," Wheeler said, "and we kept the nutrient water film very thin."[28]

They did very well, as do many crops grown this way, according to Wheeler. But traveling in a spacecraft to another planet will put constraints on the quantity and weight of commodities that could be brought along. You can't pack everything you need for a long-duration spaceflight. Some resources will need to be recycled, acquired, or produced at the destination, a process called in-situ resource utilization.

Plants could be grown hydroponically in a protected environment on deep space trips. Later, on the surface of the planet, Martian soils could be used as the growing systems need to expand. Out in space, there is no direct sunlight for plant growth, so sufficient light has to be provided artificially such as high intensity and efficient LED lights used on the ISS. It is estimated that Mars receives less than 50 % of the sunlight that Earth does, but many areas receive much less light due to low latitudes. In addition, Mars frequently has dust storms that could block sunlight for a long period of time.

In addition to all of the other challenges, food crops would most likely need to be protected from ultraviolet radiation, expanding chamber requirements to be pressurized with adequate nutrients and appropriate lighting along with and protection against extreme temperatures.

Food needs to be regenerated, or obviously it will be depleted over time. Potato tubers that contain at least one bud or eye can be replanted and produce new plants. As plants grow, they would utilize carbon dioxide and generate oxygen through photosynthesis, making the environment better for humans.

---

[27] Herridge, Linda. NASA plant researchers explore question of deep-space food crops. 17 Feb 2016. [Internet] NASA.gov; [cited 2016 Mar 15]. Available from: http://www.nasa.gov/feature/nasa-plant-researchers-explore-question-of-deep-space-food-crops

[28] Herridge, Linda. NASA.gov. NASA plant researchers explore question of deep-space food crops. 17 Feb 2016. [Internet] NASA.gov; [cited 2016 Mar 15]. Available from: http://www.nasa.gov/feature/nasa-plant-researchers-explore-question-of-deep-space-food-crops

And, as odd as it sounds, using wastewater, or even urine, as a source of nutrients for plant growth could be an option. Aboard the ISS, US astronauts use the Environmental Control and Life Support System—a system that collects and recycles used water, wastewater, and urine.[29]

Water: A Precious Resource

Water is necessary to sustain life. It is also very heavy and expensive to transport. Early manned space missions lasted hours or a few days. Now, ISS missions can last months, and it can take 2 years for a mission to Mars. Humans generate a lot of waste, which turns out to be one of the most valuable commodities in space. It can be recycled and turned into wastewater, carbon dioxide, organic solid waste, and heat. For long-term missions, it is beneficial and economical to completely recycle wastewater. Since 2009, the ISS has had a water recycling system on the level of approximately 85 % water recovery. The system uses distillation technology to treat urine, flush water, and condensate water to generate usable water.[30]

A mission to Mars would be unable to carry resupply goods such as water filters, and it is expensive to bring along spare parts to fix any possible equipment failures. NASA is currently researching a variety of water recycling technologies to adapt for use on long-term missions.

One example of a solution is the Sabatier system, originally developed by the Nobel Prize-winning chemist Paul Sabatier in the early 1900s. The process produces water and methane from carbon dioxide and hydrogen, both byproducts of life-supporting systems on the ISS. This system has been used to provide water to the ISS at the critical stage after the shuttle was retired.[31]

Resources recycled while traveling to a celestial body only addresses part of the problem for a manned exploration or colonization effort. It is a bonus if water or other resources are accessible at the destination. As an example, it was discovered that liquid water seems to flow intermittently currently on Mars. New data from NASA's Mars reconnaissance Orbiter is providing strong evi-

---

[29] Herridge, Linda. NASA.gov. NASA plant researchers explore question of deep-space food crops. 17 Feb 2016. [Internet] NASA.gov; [cited 2016 Mar 15]. Available from: http://www.nasa.gov/feature/nasa-plant-researchers-explore-question-of-deep-space-food-crops

[30] Water-recycling technology in space evolves. 10 Apr 2013. [Internet] news.weg.org; [cited 2016 Mar 15]. Available from: http://news.wef.org/water-recycling-technology-in-space-evolves/

[31] Nasa.gov. The Sabatier system: producing water on the space station. 12 May 2011. [Internet] NASA.gov; [cited 2016 Mar 15]. Available from: http://www.nasa.gov/mission_pages/station/research/news/sabatier.html

dence of streaks on slopes at several locations that appear to change over time. New findings from NASA's Mars Reconnaissance Orbiter provide strong evidence that liquid water flows intermittently currently on Mars.

"Our quest on Mars has been to 'follow the water' in our search for life in the universe, and now we have convincing science that validates what we've long suspected," said John Grunsfeld, astronaut and associate administrator of NASA's Science Mission Directorate in Washington. "This is a significant development, as it appears to confirm that water—albeit briny—is flowing today on the surface of Mars."[32]

If that water is accessible, it can be used for making propellant, sustaining human life, and growing crops. The water will not be pure and will be "salty" and contain perchlorates, and other impurities known to exist on Mars, so it would need to be purified before use.[33]

New Technology and the Road to Deep Space

One of the most exciting aspects of the future of space exploration is the cutting-edge aspect of the technology required to successfully travel into deep space. These technologies include propulsion, communication, guidance and navigation, and the recycling of resources to sustain life. Even in *The Martian*, the astronaut's survival was dependent on his creativity and expertise in order to solve problems that were not foreseen. He could utilize existing equipment but had to apply it in a unique way. This is a statement about the power of a human being to solve problems. Almost certainly, a robot could not have conducted as many sophisticated tasks as those required in this science fiction survival story.

Clearly, there are places for both human beings and high tech equipment, including complex robots, in the quest for deep space exploration. The requirements defined for successful missions to outer space and possible colonization of another celestial body include demonstrations of scientific concepts and use of materials possibly not yet discovered. Many of these efforts will be conducted through a combination of NASA, private enterprise, and international partnerships. Speaking for space enthusiasts, we are impatient to see the results.

---

[32] Nasa.gov. NASA confirms evidence that liquid water flows on today's Mars. Release 15-195. 28 Sep 2015. [Internet] NASA.gov; [cited 2016 Mar 16]. Available from: http://www.nasa.gov/press-release/nasa-confirms-evidence-that-liquid-water-flows-on-today-s-mars

[33] Nasa.gov. NASA confirms evidence that liquid water flows on today's Mars. Release 15-195. 28 Sep 2015. [Internet] NASA.gov; [cited 2016 Mar 16]. Available from:http://www.nasa.gov/press-release/nasa-confirms-evidence-that-liquid-water-flows-on-today-s-mars

Index

L. Dawson, *The Politics and Perils of Space Exploration*, Springer Praxis Books, DOI 10.1007/978-3-319-38813-7